INTELLIGENT TRANSPORT SYSTEMS IN EUROPE
Opportunities for Future Research

INTELLIGENT TRANSPORT SYSTEMS IN EUROPE
Opportunities for Future Research

Mike McDonald
University of Southampton, UK

Hartmut Keller
TRANSVER, Germany

Job Klijnhout
Rijkswaterstaat, The Netherlands

Vito Mauro
MIZAR, Italy

Richard Hall
University of Southampton, UK

Angela Spence
MIZAR, Italy

Christoph Hecht
TRANSVER, Germany

Oliver Fakler
TRANSVER, Germany

World Scientific

NEW JERSEY • LONDON • SINGAPORE • BEIJING • SHANGHAI • HONG KONG • TAIPEI • CHENNAI

Published by

World Scientific Publishing Co. Pte. Ltd.

5 Toh Tuck Link, Singapore 596224

USA office: 27 Warren Street, Suite 401-402, Hackensack, NJ 07601

UK office: 57 Shelton Street, Covent Garden, London WC2H 9HE

British Library Cataloguing-in-Publication Data
A catalogue record for this book is available from the British Library.

INTELLIGENT TRANSPORT SYSTEMS IN EUROPE
Opportunities for Future Research

ISBN 981-270-082-X

Printed in Singapore by World Scientific Printers (S) Pte Ltd

Acknowledgements

The authors wish to thank the European Commission, Directorate-General Information Society, for enabling and encouraging the 5th Framework Programme support action ROSETTA, which gave reason and resources to compile this book.

The authors wish to dedicate their work to

Fotis Karamitsos and André Vits,

David Callahan, Tony Barbas and Emilio Dávila-Gonzalez

Acknowledgements

[illegible]

[illegible]

[illegible]

Preface

This book has been written on the basis of the work done between 1999 and 2004 as part of the European Commission funded ROSETTA project.

The project was funded by the Directorate-General Information Society to identify research and other actions needed to progress appropriate applications of transport technologies in Europe. This required state-of-the-art reviews. With the aid of over 100 experts, the project team identified areas where actions were needed, the scale and character of those actions, and their subsequent promotion through a range of initiatives. This book provides insights into the Intelligent Transport System (ITS) areas identified, issues which need to be addressed and visions of what the future might hold.

It is to be noted that the book represents the views of the authors and not necessarily those of the European Commission or the ROSETTA Expert Group, either individually or collectively. Experts who have made substantial personal contributions to the ROSETTA process include Professor George Giannopoulos (section 6.1), Robert Tremlett (section 7.2), Alan Stevens (section 4.3) and Malcolm Williams (section 4.2). More generally, the authors would like to thank all the Experts who contributed to discussions and the development of understandings in ROSETTA which have formed a platform for this book.

Mike McDonald, Richard Hall, TRG, University of Southampton
Hartmut Keller, Christoph Hecht, Oliver Fakler, TRANSVER, Munich
Job Klijnhout, Rijkswaterstaat, Rotterdam
Vito Mauro, Angela Spence, MIZAR, Turin

Contents

Chapter 1

Introduction

Levels of traffic congestion, environmental pollution and safety are becoming increasingly unacceptable to a substantial proportion of the population of Europe as well as in many other developed and less developed regions. The contribution of transport to global warming is of particular concern. At the same time, our societies cannot function without adequate provision of transport to serve both the needs and desires of individuals and essential business purposes.

The introduction of new infrastructure is important but it is very clear that the construction of new roads will result in the generation of additional traffic and, of themselves, will not necessarily lead to sustainable future transport situations. Thus the general thrust of transport policy in European countries is to build essential capacity only and to better manage all available capacity so as to meet increasingly wide ranging policy objectives as effectively as possible. These policy objectives relate to curbing congestion, improving safety, addressing local and global environmental concerns and meeting broader social needs of access and mobility. The appropriate involvement of both public and private organisations and their relative roles and responsibilities is also an issue, particularly related to the financing of new and improved infrastructure and services.

The rapid developments of new technologies in the areas of location, communication, information, sensors and control are providing, and will continue to provide, ways to better achieve current policy objectives and to enable the evolution of new policies. Intelligent Transport Systems (ITS) is the collective title given to such technologies. This book deals with the applications and opportunities which are available to address specific issues of land transport using Intelligent Transport Systems and Services. Transport visions which incorporate ITS, the detail of their applications and the ways in which they may be taken forward with ITS are developed. An outline of what is considered in the book is given in Figure 1 below.

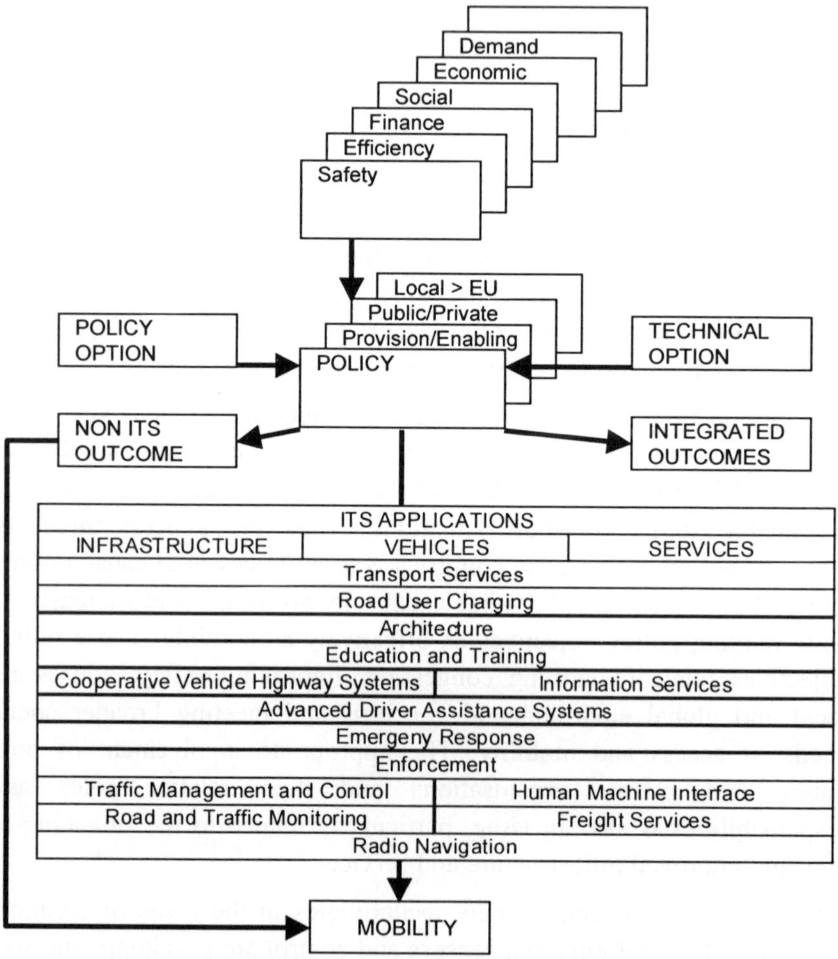

Figure 1: Outline of context

All the ITS applications addressed in the book are targeted at policy objectives and contribute to the performance of the transport infrastructure, the public and private vehicles which use the infrastructure, and a range of systems and services, many of which are only enabled by ITS. Some ITS applications are market driven and some, such as the personal delivery of online information or driver support, provide unique opportunities to integrate the use of infrastructure, vehicles and services to provide new levels of mobility and safety. Global environmental issues are of increasing importance. Some sections deal with fundamental issues such as architecture which provide essential underpinning to ITS applications.

The areas of ITS which are addressed in the book provide a vision for their application in the context of the current state-of-the-art, key issues to be addressed and the future opportunities. An initial chapter deals with the European context and a final chapter draws together conclusions and recommendations and considers issues of ITS which relate to cross area applications, such as functional architecture and education and training. Each section has been drawn together by reviews, and the opinions of the Expert Groups. ITS is a very wide subject area and, whilst the depth of considerations are similar in each section, the maturity of the subjects considered in the individual sections is very different. This has led to some differences in style, structure and referencing.

Chapter 2

Context

2.1 Getting the Benefits from ITS

This book, an outcome of the ROSETTA project, will provide the Intelligent Transport Systems (ITS) industry, the research community, central and local government and transport operators with up-to-date insights into current ITS research and applications. This will enable decision makers and others to move towards applications with a greater understanding and confidence, and to promote the research and development necessary for future deployment.

The application of transport telematics technology can revolutionise the way that people and goods move, reducing travel times, operating costs and environmental impacts. The limits of ITS are often seen by policy makers as provision of real-time information and personal travel services, traffic signal control, and highway monitoring to manage congestion and allocate priority on highway networks. However, ITS have far wider applicability.

In traveller services, ITS can facilitate ticketing and intermodal transfer for passengers, as well as giving them confidence and information throughout a trip. ITS also have huge potential to improve the operation and performance of individual modes (ROSETTA, 2004 (1) & (7)). For example, modern Demand Responsive Transport (DRT) is dependent upon ITS, with vehicle and passenger routeing and operating efficiency transformed by ITS control. In goods transport (ROSETTA, 2004 (10)), new technologies and transport telematics can ensure optimum use of individual freight modes, allow modes to be combined efficiently and remove or reduce paperwork, with electronic documentation, identification of consignments and customs control. In private transport, vehicle-highway communications (ROSETTA, 2004 (8) & (9)) can reduce journey times and increase reliability for the user, and can help to manage supply through mitigating the effects of limited road space, improving the efficiency of network operations and minimising disruption from maintenance, renewal or traffic incidents.

Some of the main applications of ITS will be in increasing the safety of travel (ROSETTA, 2004, (4)). International and local security measures can be facilitated by ITS; ITS in railways helps to prevent incidents in normal running and under maintenance. On the highway (ROSETTA, 2004 (4), (6) & (8)), automated enforcement of speed and signal observation reduces collisions and their consequences; ramp metering can reduce congestion and conflict on busy highways; road traffic monitoring can assist in crisis planning and management. Also, ITS can make a great difference to emergency services (ROSETTA, 2004 (2)). It can provide priority for emergency vehicles on congested networks, and can offer open systems architectures to accommodate calls and responses relating to radiation, chemical and oil spillage, road or rail incidents, forest fires, mountain and cliff rescue, maritime search or rescue and even domestic emergencies, such as to an elderly person living alone and requiring medical help.

Conditions must be provided which facilitate ITS development, adoption and efficient operation if the maximum potential is to be achieved. The general adoption of the European ITS framework architecture will greatly assist the co-ordination and interoperability of ITS (ROSETTA, 2004 (12)). The key to unlocking the potential benefits of ITS in traveller services is standardisation of data requirements and data management, which will enable intermodal management and inter-operator information exchange (ROSETTA, 2004 (various)). Policy makers can facilitate this by agreeing data protocols, performance standards and standardised testing methods for new systems and equipment. The research industry can contribute greatly to the development of standards, including reporting in a more consistent framework. Investment decisions will be helped by research into the equity and welfare effects of ITS.

The last part of this chapter looks at what needs to be done to create the conditions in which ITS can contribute to improving the efficiency and reducing the environmental impact of transport. The importance of ITS to the future of Europe stems from the scale of transport activities and the related costs, time, and social and environmental impacts. ITS will bring more choice to all those involved in travel and transport. If choices are made wisely with full appreciation of the consequences, individuals and society will benefit with efficient, safe, socially inclusive and sustainable transport which will contribute to a healthy European economy. Key to the future will be to understand the behavioural response to the transformation which ITS will bring, and several of ROSETTA's work areas have explicitly identified such needs.

2.2 Transport in the EU

In 2002 (European Commission Eurostat, 2004), freight transport, excluding maritime, amounted to 2,158 billion tkm (tonne kilometres), or 4,764 tkm per person. Private travel, excluding air which is growing rapidly, totalled 5092 billion pkm (passenger kilometres), which averaged 11,240 pkm per person. Private households spent 14 % of their total expenditure on transport. These figures indicate the scale of the economic, social and environmental significance of transport in Europe.

2.2.1 *Accidents*

– The number of road accident victims remains high in the European Union, with around 40,000 fatalities and 1.7 million injuries a year, although safety has improved in recent years. The directly measurable costs of road accidents are of the order of € 45 billion, with indirect costs three to four times higher.

- There are significant differences in accident rates between Member States, with some recording more than two and a half times as many deaths per head of population than others, indicating that there is considerable room for improvement (European Commission Eurostat, 2004). The safest roads are found in Sweden and the UK, which in 2003 recorded rates of 5.9 and 6.1 road deaths per 100,000 population respectively. The highest death rates are found in Poland, Portugal and Greece where car ownership has grown rapidly. In 2003 the death rates were 14.8 road deaths per 100,000 population for Poland and Portugal and 14.7 for Greece.

2.2.2 *Congestion*

- Today 79 % of passenger transport, and 44 % of freight transport, is by road. Congestion is widespread throughout the EU. Many European Member States predict increases in urban motorway travel of some 50 % by the end of 2005. Growth of this magnitude will lead to increases in delays of 400 %, unless traffic is managed more efficiently (McDonald et al., 2000). The road share of goods traffic has been growing and will reach 47 % by 2010 if no action is taken.
- The trans-European transport network suffers increasingly from chronic congestion: some 75,000 km, i.e. 10 % of the road network, is affected daily by traffic jams. The external costs of road traffic congestion already amount to 0.5 % of the EU's GDP. If nothing is done, road congestion will increase significantly by 2010 and beyond (Yrjo-Koskinen, 2002).
- Congestion does not just affect roads. Across the EU 20 % of the rail network (16,000 km of railway) is classified as being bottlenecked, and 16 of the EU's main airports record delays of more than 15 minutes on 30 % of their flights (European Commission, 2001 (1)).

2.2.3 *Environmental costs*

- Energy consumption by the transport sector has grown at a faster rate than that of any other sector, resulting in an increasing share of energy consumption over time. By the late 1990s, the transport sector accounted for a larger share of total energy consumption than the industrial sector. Within the transport sector, road transport is by far the largest energy consumer, accounting for about 83 % of total transport energy demand (European Commission Eurostat, 2004).
- While considerable progress is being made in other sectors, CO_2 production within the transport sector has risen steadily, as the volume of road traffic increases.

2.2.4 *Social costs*

- Road traffic in general and cars in particular are the main source of urban air pollution, and urban air quality remains poor in many cities (Lynham, 1997). Northern Europe is generally less polluted than Western and Southern Europe because of lower levels of car use, but even in the cleanest cities, people are exposed to levels of pollutants that can have adverse effects on health. This has significant implications for human health within the EU because 70 % of the EU population lives in urban areas (Lynham, 1997).
- Another significant impact of road traffic is noise. Road traffic is the main cause of noise disturbance and has emerged in recent years as an ever present but often underestimated pollutant in urban areas. In Europe, it is estimated that 20 % of inhabitants suffer unacceptable levels of noise pollution from road traffic (Lynham, 1997).
- Across Europe, car mobility is also impacting on health by reinforcing an increasingly sedentary lifestyle and contributing to a decrease in the modal share of healthier walking and cycling.

- Within the EU, 27 % of households do not own a car. Households without access to a car are reliant upon public transport, yet the decline in growth of most public transport services is threatening the economic viability of these services. There is a risk that the non car-owning sector may suffer social exclusion if service provision is reduced.

2.2.5 *Economic costs*

- The cost of these social, environmental and economic impacts to society is very high. Congestion costs the European Union about 0.5 % of its annual GDP, accidents cost 1.5 % and pollution and noise cost at least 0.6 %. The total cost is about € 250 billion, with 90 % of this cost being attributable to road transport (European Commission Transport website).
- These costs will continue to rise as demand for travel increases and the EU Commission has identified congestion as a serious threat to the economic competitiveness of the EU (European Commission 2001 (1)).
- At present these 'costs to society' are not borne directly by the people responsible for them but are externalised, and therefore not considered when individuals make their transport choices. This is true both for personal transport choices and for corporate freight transport choices. However, both personal and freight transport users remain convinced that the costs that they are required to pay, particularly for road use, are too high, even though the major societal costs are excluded.

2.3 Transport Policy

Transport has always featured highly in EU policy. Transport policy objectives initially focused on removing checks and formalities at borders between Member States to ensure freedom of movement for people and goods within the Common Market, based on a two-fold approach of liberalisation and harmonisation.

As progress on the single market continued, and with the enlargement of the EU, it became apparent that some problems could not be resolved by liberalisation and harmonisation alone. The creation of a single market has given a great boost to the growth of transport in Europe, but the pattern of this growth has been uneven, favouring car passenger transport and road freight transport over other modes. The results of this imbalance in growth are evident as road congestion, pollution and accidents, which act against sustainable development and the opportunities inherent in the single market. The situation has been further aggravated in many instances by a lack of vision and overall view of transport, so that the different modes of transport have each evolved in an isolated, non-complementary fashion. The Commission has recognised that an integrated transport policy is of crucial importance to the efficient functioning of the economy and to the mobility of people and of goods (European Commission, 2003 (3)).

The EU White Paper (European Commission, 2001(1)) on transport has the following main objectives:

- Shifting the balance between modes of transport by 2010 by revitalising railways and promoting maritime and inland waterway transport;
- Achieving fair price systems which reflect the true costs of transport, including external costs such as environmental damage, congestion, or human accidents;
- Making transport systems more efficient and safer.

The sustainable mobility approach shapes a number of specific objectives now being pursued within Europe (see Table 1 below).

Table 1: Sustainable mobility approach – objectives

GOAL	REASONING	APPROACH
Intermodality	Transport systems must be developed in which different modes of transport complement one another, so that passengers and freight use the form of transport that is the most efficient and best suited to the purpose at each stage of the journey. This will support optimum mobility and a more balanced distribution of traffic.	Develop efficient interfaces between modes. The establishment of Trans-European Networks for road, rail and air, so that all parts of the Union can be interconnected, is a major project within this area.
Improved access to public transport for passenger and freight travel	Promote easily available, efficient, safe and affordable local public transport services. Public transport is the only form of transport available to all citizens, particularly in large cities. It plays a key role in social development, particularly by improving accessibility and the situation of weaker regions and disadvantaged social groups.	Encourage use and development of public transport, bicycles and walking in urban and wider urban areas. Improve access to public transport and improve integration between modes.
Mitigation of the impacts of road transport	Increasing demand for road transport is unsustainable in relation to its environmental, social and economic impacts. The impact of road transport must be mitigated as much as possible.	Develop more efficient use of road infrastructure through application of ITS, to manage congestion and pollution, and provide an alternative to road building. Manage demand for road transport where there are public transport alternatives, through mechanisms including road user charging and improved land use planning. Develop advanced transport energy technologies which cut the consumption of oil and gas. Develop hydrogen-powered zero emission vehicles. Develop the potential of alternative fuel technologies. Promote road safety.

GOAL	REASONING	APPROACH
Increased modal share by public transport	Increasing the modal share of public transport will also contribute to environmental and safety objectives. Promote rail, maritime transport, inland waterways transport and intermodal transport for freight.	Provide mechanisms for financial support for these modes and internalise the external costs of road transport.
Make better use of existing networks	Optimising efficient use of existing networks will support sustainable transport objectives.	Application of advanced technologies, Intelligent Transport Systems and development of the European Galileo satellite navigation system.

2.3.1 *Forward trends*

The recent enlargement of the European Union with 10 new Member States has increased the EU land area by 25 % and population by 29 %. In 1998, exports from the new Member States to the Union were already running at 112 million tonnes, 2.2 times the 1990 level. Enlargement will speed up the increase in traffic still further, notably for freight transport. New transport corridors will be built in the 10 new EU Member States in order to create a transport network which more effectively links to the current 15 members. It is estimated that a total of 20,000 kilometres of road and 30,000 kilometres of railway lines need to be built, plus new airports (European Commission, 2003 (1)). Transport rose by around a third from 1990 to 2002, broadly in line with GDP growth. The expected growth of GDP in the coming years is higher, as the accession states 'catch up'.

In freight, there are a number of existing and new pressures which will tend towards the take-up of ITS. The existing trends include the wish to reduce internal and external costs and the desire to move to a logistics system which is more intermodal, paper-free and real-time, with better tracking, tracing and security. The new trends include the globalisation of production, distribution, transport operators, services and clients. These have led to a persistent increase in freight volumes and to a shift in the pattern of distribution, with many more delivery points arising through e-commerce.

In passenger transport, there are factors working both for and against the adoption of ITS. Individual service operators, routes and terminals are motivated to apply ITS, because it provides a better product. The essential requirement for a seamless journey is, however, seamless information and management – and the competition between operators has tended to reduce joint working (ILS NRW et al., 2005).

Another reason why ITS implementation is important is that Road User Charging (RUC) or Value Pricing (VP) cannot easily be carried through without ITS. Simple, local systems may be possible, but without ITS it will not be not possible to carry out the EC's commitment to road user charging (European Commission, 2003 (2); Glaister and Ochieng, 2003). Road pricing has the potential to hit many transport targets: improving equity, reducing environmental damage and raising revenues for public transport investment. As part of this, road pricing will encourage the choice of the right mode for the journey. 'Public Transport good; Cars bad' is as much a nonsense as any other statement of such extremes. Where a lot of people are going from roughly the same origin to roughly the same destination, public transport has lower external costs and the financial incentives should be planned to encourage that choice. Where there is low demand for a single journey, and where walking or cycling is impractical, the car is the best mode in external terms as well as internal: running hourly bus services on convoluted routes to carry a tiny number of people is neither cost effective nor environmentally sound. Glaister and Graham (2003) point out that a national system of road user charging in the UK would reduce the cost of driving in rural areas, bringing significant net benefits.

2.4 Socio-economic Trends

Current socio-economic changes mean that new travel demands and trip making characteristics are emerging. This makes it increasingly difficult to deliver an attractive, efficient and sustainable transport system.

2.4.1 *Increasing life expectancy – ageing population*

- The proportion of the population aged over 65 has increased within the EU as life expectancy is increasing with improved medical care. In 1990, people aged over 65 years accounted for 14.5 % of the population of the EU, and by 2000 accounted for nearly 16 %.
- This trend is set to continue across the EU and within the UK it is forecast that the percentage of the population aged 65 years and over will increase from 15.8 % in 1998 to 19.2 % in 2021. A 'Help The Aged' report forecasts that by 2021 one in three people in the UK will be aged over 60 (DETR, 2001).
- The transport requirements of a growing population of older people must be identified and met. For example, older people place a difficult demand upon public transport services because those without access to a car cannot be completely supported by a 'hub and spoke' public transport system; they are likely to require a door to door public transport service due to reduced mobility, in order to mitigate social exclusion.

2.4.2 *Falling household size*

- Average household size has diminished within the EU over the last 20 years. In the UK the average household size was 2.7 in 1981 and 2.3 in 1998, and is predicted to be 2.15 by 2021. Within the EU the average household size was 2.8 in 1981 and 2.5 in 1998, although this figure varies between Member States; average household sizes within Mediterranean countries and Ireland tend to be larger than household sizes in continental and Northern Europe.
- The trend for smaller households is compounding the problem of housing the growing population. The implication of low-density suburban housing is that there will be a high demand for car travel because low density cannot easily support such a dispersed population. Instead more flexible public transport

services will be required. High density urban housing can be served by public transport much more successfully.

- There is a high expectation that an integrative view of land use and transport planning will be needed and will help in the future.

2.4.3 *Population dispersal*

- In Europe people are moving out of cities. This process of decentralisation of cities has been facilitated by the car and has resulted in dispersed travel patterns, reducing the possibility of promoting efficient public transport (Banister, 1999).

2.5 IT and Opportunities for Changes

Current reliance upon the private car for passenger transport, and road vehicles for freight transport in Europe is undermining sustainable transport goals. In addition, social and economic trends ensure that the goal of sustainable transport continues to be elusive, as user needs become more complex. Innovative approaches to the ways in which people and goods move are needed, and deployment of technological solutions are seen to be the key to progress.

Information and related technologies will be critical components of future travel and transport solutions, where the focus in on sustainable mobility and addressing changing user needs. The ITS applications that are evolving cover a broad scope of information and telecommunications technologies that detect people, drivers, vehicles, goods, traffic and environmental conditions, and communicate this information to a variety of end users.

Since the provision of information about the impact of transport choices and the collection of the true costs associated with them are key elements in encouraging more sustainable transport choices, and both of these aspects are central to the remit of ITS, directing the use of the developing technology in this way should be a main component of transport policy.

2.5.1 *Supporting sustainable mobility objectives and user needs*

ITS are able to support sustainable mobility by providing opportunities to make better use of existing networks, to manage demand, to improve safety and to support high quality, flexible passenger and freight services (see Table 2 and Table 3 below).

Table 2: Sustainable mobility through traffic information and control

TRAFFIC INFORMATION AND CONTROL	ITS APPLICATIONS	ROLE IN SUSTAINABLE MOBILITY / USER NEEDS
Vehicle control	Driver stop/rest alert. Driver collision warnings (front, rear, side, pedestrian, obstacle), emergency control intervention to avoid collisions. Environmental condition warnings such as fog or ice, vision enhancement in adverse driving conditions. Vehicle speed control (Intelligent Speed Adaptation), Low Speed Automation for lower speeds and stop & start/ congested conditions, Automatic Cruise Control (ACC), automatic lane merge/overtaking manoeuvres. Route guidance, electronic vehicle guidance on segregated and non-segregated routes which includes areas such as bus ways and freight terminals, electronic tow-bar and platooning, lateral and longitudinal control.	Overall enhancement of safety and driver comfort. Improved environmental performance of vehicles through smoother driving. Improved network management through route guidance information. Improved public transport performance and fleet management through vehicle guidance and freight platooning.
Traffic control	Detection of traffic conditions for ramp metering, VMS, real-time signalling.	Improved network efficiency and safety, reduced congestion and associated fuel consumption and pollution. Provides real-time location of vehicles for public transport priority.

TRAFFIC INFORMATION AND CONTROL	ITS APPLICATIONS	ROLE IN SUSTAINABLE MOBILITY / USER NEEDS
Pre-trip and On-trip information	Pre-trip and on-trip multimodal traveller information through a variety of interfaces, provision of parameter-specific information and personalised travel information.	Improved travel information, improved network management as users make more informed travel choices, promotion of public transport and intermodality.

Table 3: Sustainable mobility through management services

MANAGEMENT SERVICES	ITS APPLICATIONS	ROLE IN SUSTAINABLE MOBILITY AND USER NEEDS
Passenger transport services	Demand responsive transport Seamless ticketing	Flexible transport Promoting public transport Improved knowledge of public transport user behaviour for improved management
Freight transport services	Monitoring of freight container conditions	Improved fleet management
Emergency service systems	In-vehicle Mayday system	Improved emergency response and safety
Enforcement	Toll collection technologies, CCTV, etc.	Supports a variety of measures
Mayday services	Emergency response teams – rapid dissemination of detailed accident information.	Improved safety
Demand management	Seamless payments for tickets and tolls, support of physical demand management strategies	Demand management and Improved network efficiency

2.5.1.1 *Good decisions in Europe*

Decision making in the European Union is arranged to be relatively straightforward: the Commission presents a proposal, usually after consultation; the European Parliament adopts or revises the proposal (either by qualified majority voting or by unanimous decision, depending on the subject of the proposal – for example, taxation questions require unanimity); the European Council then gives its assent.

This process was evolved to serve a community of initially six nations. With 25 nations, different decision making processes are being sought (European Commission, 2003 (4)). The Treaty of Nice reduced the areas of policy which require a unanimous vote and reduced the number of Members of the European Parliament from existing Member States, in order to accommodate the enlargement of Europe from 15 to 25 (and later 30) members (European Commission, 2001 (2)). The proposed European Constitution which enshrines these changes also proposes to extend the EU's scope of action in several policy areas (e.g. foreign policy).

In this state of flux, it is worth bearing in mind that two principles operate most strongly in policy areas such as transport. First the motivating principle in the European Union is to serve 'the interests its peoples and nations share together' (Fontaine, 2003). That is likely to be the foundation of the future decision-making process, whatever it is. ITS are part of that common interest: increasing trade and social mobility, reducing environmental damage and reducing the costs of transport.

The second, and most critical, principle is that of subsidiarity. The EU is 'subsidiary' to Member States and can only act in areas where 'the objectives of the intended action cannot be sufficiently achieved by the Member States but can... be better achieved at Union level.' The fundamental principle here is that the powers of the EU are derived from its members, rather than the other way round. In the proposed amendments to the European Constitution, subsidiarity would be extended explicitly to local and regional decision making.

It will be necessary for those seeking decisions in ITS to recognise the complexity inherent in decision making given the different cultures and different policy priorities of the 25 nations which make up the European Union (Stevens, 2003). At a day-to-day level, policy makers recruited or seconded from the Member States make up the decision-making engine of Europe (Stevens and Stevens, 2000). It cannot readily be assumed that there is a 'right answer' to any policy question, since the needs, capabilities and policies of different countries may be different at any one time and may change over time, often in different directions. It may be more useful to follow the OECD 'good decision' model (OECD,

2002), and aim to develop ITS solutions which can be fitted to a range of policy environments. In ITS terms, that mainly means developing *interoperable* rather than *common* systems and rules. This approach recognises the diversity of interests as well of delivery cultures in Europe and also the practical reality of 'subsidiarity': although it is important that ITS be applied on a trans-national scale, the decision-making reality is that most transport decisions are, and will remain, national decisions.

2.5.1.2 *ITS research*

ITS is a rapidly changing subject area. If transport telematics are to be exploited in Europe to their full potential, decision makers and potential users must be kept informed of current thinking and knowledge. The ROSETTA process has facilitated this, first by gathering the latest documentation and information into one source, then by exposing that information to critical examination and finally by reporting on the state-of-the-art and the critical issues.

2.5.1.3 *The future*

Technology is available which can locate, identify, provide access, charge, communicate, support and manage people, goods and vehicles in a range of environments. At present, some of these technologies are expensive, inaccurate, unreliable, have limited functionality, lack integration and are intrusive. However, such technologies and related systems are being developed and improved at a very rapid rate. Some are the result of technology push rather than user pull, although the latter is increasingly driving ITS forward.

A crucial factor for the early deployment of ITS is enhancement of the way in which humans are involved, i.e. the more rapid development of Human-Machine Interfaces (HMI) which give professionals and the public confidence in using ITS. The Internet has done a great deal to increase people's confidence in getting information out of Information and Communication Technologies (ICT), but there are great strides still to be made before this is universal and before it extends to confidence in the information terminals on streets and at transport nodes, and for other applications.

Effective HMI must be based on understanding the needs of users and the process of diffusion of new systems into the market, as well as the business processes which determine the flow and structure of information. HMI must accommodate future needs and wants arising from developments in society, such as the needs of different user groups, the special needs of elderly people, disabled people and those with learning or language difficulties.

For Advanced Driver Assistance Systems (ADAS), a critical HMI factor is that the system should not compromise driver safety. This is most important 'during high task workload', that is, when the driver is under stress. This is a high aspiration given that many ITS systems are used only at times of stress – I'm lost; I need a parking space; I don't know how to pay to cross this bridge and I'm late and I have to get across this bridge. It is compounded by the fact that different people have different stress levels, different perceptions of risk and different competence in driving and in taking and following instructions. However, there is a strong market driver for manufacturers to overcome these difficulties with the world market for in-vehicle driver support systems booming and much commercial and other research underway. This is driving a related market for services extending to the office and entertainment, as well as way-finding, information and logistics and fleet management.

For travellers, an ITS system should deliver information in a straightforward and clear way which reflects the needs of the individual. This is a very complex process which involves understanding time, cost, and physical constraints of the individuals or groups concerned. Much research is required to provide systems and services which meet these needs as well as all the ITS technologies and systems to assemble and manage the essential databases of underlying information. This requires intermodal real-time electronic information and transaction systems in passenger and freight services and further harmonisation of message and document standards in telematics. There are great opportunities for saving costs, time and environmental impact through the expansion of e-commerce, leading to more efficient freight movements and, potentially, to a lower number of passenger journeys.

The most important actions for Europe are policy actions. Whilst the market has provided a great many ITS products and developments, for ITS to have their full impact on a trans-national scale the integration of the many valuable, but still separate, ITS initiatives is essential. EU research funding has been generous, but it is now more critical that policy developments enable new co-working approaches, some of which may need to take place outside normal competition rules. Businesses are jealous of information: they invest a great deal in developing ways of collecting and presenting information to customers to get a market advantage and to develop internal management communications that make their operations more efficient. Therefore, a mechanism for integration will be difficult as they are reluctant to share internal and external information content or format. Policy makers will have to provide the incentives to overcome these problems. However, there is a subsidiary point that competition rules can make it difficult to share information without appearing to form a cartel (ILS NRW et al., 2005).

Other important requirements before ITS can fulfil their potential to transform transport in the EU include the need for Europe-wide open standard ITS-based tracking and tracing systems, to facilitate intermodal transfer and to give consigners confidence in security. In the EU context, security and enforcement will require cross-border procedures for pursuing offenders, along with improved protocols for data standards and communications management.

The range of vehicle-highway communications for safety is expanding rapidly and the take-up of systems by vehicle manufacturers and purchasers is reaching the point at which highway owners need to determine their strategies for speed and collision control. A balance between systems in which the vehicle is controlled by its own understanding of the local environment, and systems in which the network owner takes control, must be agreed. Leaving development to the vehicle manufacturers is likely to lead to the adoption of the former approach, though there are likely to be much greater safety and efficiency benefits of the latter. Such decisions should be evidence-based, with the benefits as well as the costs of a managed system well understood before a decision is taken *not* to invest in it. Research requirements are for assessment methodologies to understand the costs and benefits of driver assistance through vehicle-highway communication; research on the most suitable technologies and systems architecture for large scale applications (possibly quite different from those being developed for individual areas); and a long term strategy to merge the market-driven development of systems to support the driver with the infrastructure systems required by authorities for safety, network management and charging. Improving the monitoring of highway conditions and use will also help to enable the maximisation of the benefits of ITS. However, new systems and services must address future problems, not just those which currently exist.

The final critical step in ITS research has yet to be taken: gaining an understanding of sociological and behavioural responses to ITS. What systems will and won't be accepted by which people? Will people change their lifestyles in response to ITS and to the policy instruments that they facilitate, such as road user charging? How will different people cope with the intelligent transport society? Consider a retired manual worker with poor literacy, few learning skills and failing eyesight, hearing and response times – how will he or she respond to the intelligent vehicle-highway? Consider a 19 year-old, about to start his or her degree course – will they even bother to buy a car, moving in congested networks, or will they choose intelligent public transport, joining a car club for occasional journeys? In the long run, the sociological and behavioural response will determine the take-up of systems and hence their viability politically and commercially. We have been researching ITS for a generation and the research emphasis must shift from the technology, which is fairly mature, to understanding behavioural response.

ROSETTA has drawn together the current research and practice in ITS. The insights here should help policy makers prioritise the integration and promotion of ITS implementation to transform transport. It should also encourage central policy makers and professional bodies to enhance the ITS training and education networks.

Policy makers do not have a choice as to whether ITS will 'happen'. The cost savings and time savings, the safety benefits and commercial value, the confidence and communication which ITS bring, all mean that ITS will be implemented. The policy question is whether they will be implemented piecemeal, by individual towns, cities and highway operators, by individual vehicle manufacturers and public transport operators seeking market advantage, or whether they will implemented in a coherent manner, so that the environmental and efficiency savings can be derived quickly and on the large scale, across Europe. The policy of piecemeal development was sensible in the early days of ITS as it was not known which ideas would be effective, or what the scale of the effects would be. Much of that now is known, and a different policy approach is required. A continuation of the policy of piecemeal development will have the damaging effect of keeping costs artificially high. This would impact adversely on the EU's competitive position and also keep welfare and environmental costs of transport artificially high. Moreover, it risks the development of systems which become increasingly difficult to integrate, creating an unnecessary future cost burden to move to interoperable, trans-national ITS. Only an integrated approach will ensure that Europe reaps the full benefits of its past leadership and expenditure on ITS.

Chapter 3

Traveller Services

3.1 Passenger Transport Services

The situation facing Passenger Transport Services (PTS) is a familiar one – while indispensable in urban areas, they are often barely present in scantly populated regions and constantly in competition with individual transport modes. But can this be considered an inevitable (and unchangeable) picture? Future prospects appear to confirm the continued importance of PTS. Further increase in transport activities along with the need to reduce vehicle emissions, as underlined in the 'White Paper – European transport policy for 2010: time to decide' (European Commission, 2001 (1)), imply a more important role for PTS in the future, as well as the need to develop innovative solutions.

This chapter introduces some possible 'visions' for passenger transport services, accompanied by a brief review of the current state of technology and suitable or necessary research activities. As passenger transport services embrace a very wide range of topics, this section will concentrate on central issues of PTS, namely Intermodality, Passenger Transport Systems and Operation, Payment Systems and Demand Responsive Transport (DRT). Linked topics such as Safety and Security, Monitoring, Information Services or New/Alternative Transportation Modes are highlighted in other sections of this book.

3.1.1 Vision

Developing a long-term vision on any kind of development implicitly requires a large number of assumptions on the future framework. Many developments, not only in the technology sector, are closely linked, influencing and depending on each other. The development of ITS in the transport sector is strongly linked to the overall development of transport demand and supply for the future. These, again, depend on a multitude of economic, ecological, social, political and cultural factors. Land use development needs to be considered, along with the general wealth and leisure time of people in Europe as well as their level of environmental awareness.

It is not the intention of this chapter to provide a systematic analysis of the background leading to the ideas sketched in some of the visions. This has already been done by experts in the respective fields of transportation. The assumptions we have adopted, assuming more or less a trend development for most aspects, are as follows:

- Transport demand will continue to rise in Europe due to increasing wealth and leisure time;
- Road transport will subsequently suffer from congestion, as physical infrastructure cannot expand as required;
- All modes of transport will become more environmentally friendly, less polluting and require less energy;
- Information supply will be ubiquitous and delivered by physical and wireless data networks;
- Costs for information technology will steadily decrease.

Telematics applications will improve the performance of every mode of transport available in Europe. Even more, they will help to integrate different means of transportation to enable a shift from individual transport systems to passenger transport services. So what is the difference between passenger transport services and Public Transport (PT) as known today?

Over the past 50 years we have witnessed the unprecedented success of private automobile transportation throughout Europe. It was the vehicle of industrial prosperity and individual freedom in all our countries. However, environmental impacts and increasing congestion both in dense urban areas as well as on major motorways led to the political consensus that automobile travel alone could not provide the sole solution to our transportation demand. For so many years this was the reasoning behind an ever increasing subsidy for classic public transport. But automobile travel and public transport have always remained competitors – with public transport losing ground all the time.

Only the emergence of Information Technologies (IT) services in transport provided the chance to move from an operator's view towards a customer-oriented view of transport services. Travellers do not care about operational restraints or political agendas: they want the fastest, cheapest – and especially for elderly people – the most comfortable or convenient way to reach their destination. Information technology provides the tools to integrate all the separate and competing systems to offer seamless travel to the customer, combining the best means of transportation for each part of the journey to achieve the highest performance imaginable. Hence passenger transport services imply a move from the supply side to the demand side – passenger needs are at the centre and ITS provides the means to meet them.

3.1.1.1 Intermodality – integration of modes

Here we present our vision of a perfect intermodal passenger transport system, as we can imagine it today.

A range of modes is available, each suited to specific tasks within the overall system. Modes do not compete, but complement each other as feeder and as backbone systems to form a well-organised hierarchical network that flexibly adjusts the service offered to actual travel demand. Every region of Europe is connected to this network by a system of flexible transport services and high-speed links between all major conglomerations. Public transport schedules are interactively adjusted in cases of delay to allow fast and comfortable change of modes. Different means of transport are physically connected in well-designed interchange terminals, so that transfer between modes is smooth, safe and easy. To avoid customers regarding travelling by PT as 'uncomfortable' or 'wasted time', additional services provided in-vehicle, at transfer points or on platforms will offer commercial, leisure and entertainment facilities.

Passengers do not approach individual modes, but the entire system as one entity. All modes are linked within a 'common shell', so passengers do not need to worry about mode or operations when they are planning a journey. Service points provide information, guidance, seat reservation and tickets as well as value added services for the whole intermodal transport system. Passengers may inquire about conditions for a regular commuter trip or for an intermodal journey around Europe at the same place. Information on intermodal transport and most of the services is also available in private homes and business offices or may be obtained with the help of mobile personal travel assistance devices. Reservation and ticketing for multimodal trips is well integrated. When there are delays, the booking is changed automatically.

Automated electronic payment is harmonised across Europe for all mass transit, so passengers can enter and exit public transport at will without worrying about having the right ticket. Intelligent payment systems calculate fares individually to guarantee the best tariff for the journey. This has at the same time considerable benefits for transport operators as it provides valuable information on travel demand. Data on travel patterns can be collected from booking, route planning systems, etc. to create the best possible service for customers.

An important feature of a truly intermodal transport system is the existence of multimodal, multi-operator co-operation to handle incidents. A breakdown in one system can be quickly compensated for by additional supply from associated networks, as well as routeing advice on all modes which immediately takes the incident into consideration.

Figure 2 shows the different demands and links of an intermodal transport system.

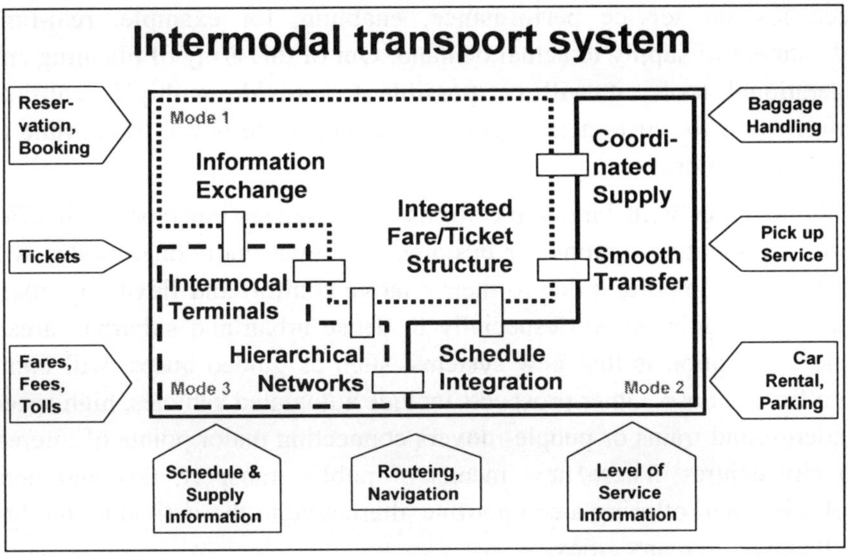

Figure 2: Intermodal transport system

3.1.1.2 Passenger transport systems and operation

The next step is to define what future telematics applications are needed to help the operators in providing the best service to travellers. This includes the operation of vehicles themselves as well as related real-time information to enable adequate utilisation. We assume that the progress in GPS/GALILEO, WAP, GPRS, UMTS (Internet access via mobile phones) or equivalent 'personal travel assistants', contactless smartcards, etc. will allow their use on a large scale (cheap, easy to use). But providing a better service to travellers has an important prerequisite – operators can only share better information if they have it themselves! The most promising concept is the integration of all available information in a data warehouse concept, via standardised interfaces and open system architecture.

Newly developed planning and operational tools will be able to process the information collected from planning processes, routeing, schedule databases, ticketing and demand request data, vehicle and driver schedules, vehicle maintenance, etc. This will also provide immediate feedback on service performance, enabling, for example, real-time adjustment of supply to actual demand. Out of this array of planning and operational tools, it will be possible to provide multiple real-time information for customers, reporting real departure times, travel time to final destination, etc.

This, together with innovative forms of passenger transport, will offer more competitive options for customers. Though individual modes will still be very important due to their relative comfort and flexibility, mass transit will gain ground especially in dense urban and suburban areas. One assumption is that new systems, such as guided buses, will come into frequent use. Other prospects include automated vehicles, high speed underground trains or people-movers connecting major points of interest in city centres. These new means of public transport, fast and non-polluting, and offering a competitive alternative to the individual modes, will appear in many cities.

As there will be several different ways of undertaking a trip, transport service disruptions caused by breakdowns, natural causes (e.g. flooding) or by human action (staff on strike, passenger problems) will be handled more easily. This will be both in terms of recovering the situations and of helping travellers to go on their way. Therefore, travellers must be guided when a part of the transport system is down, but we must also consider the transport system when a piece of the telecommunication system is down.

3.1.1.3 Payment systems

Electronic payment is not specific to transport services. Banking and commerce is pushing towards new solutions, reducing the demand for coins and notes. There are specific aspects to payment in public transport, however. Usually the customer has to decide on the ticket or fare at the beginning of the journey. Most systems behave rather inflexibly in adjusting to a later change of plans – meaning if the customer decides on a different route or means of transportation on the way, the original fare is lost. The same holds true for period or season tickets; if a traveller does not use it as expected, the cost could be higher than if paying the fare for individual trips.

ITS provides the possibility of monitoring transport use at an individual level for optimising the service and charging only at the end – of the trip or the month – at the most favourable rate applicable. This also increases comfort for all these occasional travellers not familiar with the vast variety of different tariffs most public transport providers offer nowadays. An ideal solution could comprise a single act of registration, a contactless paying system such as a smart card and a monthly transportation bill charged to the traveller's bank or credit card. Passenger transport services would monitor every use of transport services and charge the best rate applicable.

Ideally, electronic payment systems will be compatible from one transport network to the next and even across Europe. Integration with different retail and service payment systems is also desirable. Collected data may also be used for enhanced operation and planning of the linked transport services.

3.1.1.4 Demand responsive transport services

The main focus of Demand Responsive Transport (DRT) is to fill the geographical and service gaps in Public Transport (PT). The open-for-all DRTS (Demand Responsive Transport System) guarantees the equality of flexible public transport for everybody and for every direction in this area. The demand of DRTS is an essential part of creating seamless public transport for areas from the city core up to the rural outskirts.

The need of DRT is everywhere, but the type of needs vary depending on the operating environment of other PT. Therefore there is no one DRT system suitable for every location and case. The vision is that every type of transport demand could be met with the proper type of DRT system. In urban areas, where PT services are widely deployed, DRT may aim at special services for handicapped citizens or elderly people. In rural areas DRT will present a solid alternative to individual modes or for infrequently operated public lines.

But DRT should not explicitly concentrate on being the transport mode for times and zones of special or low demand. Assuming an adequate number of deployed vehicles, new concepts like Intelligent Grouping Transportation (IGT) will also provide alternatives to individual or regular public transport. Overall transportation efficiency and reduction in costs arise from the method of grouping passengers with compatible itineraries into the same minibus vehicles. A computerised system organises this grouping so that the vehicle can transport all passengers on-board from door-to-door.

Prerequisite for many companies jointly operating an area is a strong linkage of local services to create a large scale flexible transport service under a common shell that may approach the flexibility of the private car. This includes co-operation models, rules and data bases with regular public transport. Advanced systems especially developed for the needs of DRT services will manage the bookings and operation of multiple service providers to give the best service for the user at the best price. There is a great need for predictive optimisation, where the pick-up time, fare, as well as the travel time, can be given to a new passenger during the primary booking call. Another advantage will be automated booking and payment facilities.

Another facet of DRT would be transport modes that could handle freight or passengers at different times of the day, depending on demand. First experiments with tramways (Volkswagen in Dresden) are promising and could be expanded to combined delivery and passenger services in rural areas.

3.1.2 State-of-the-art

The short visionary notes above provide an expert's view on development and performance of passenger transport services in the future. The following is an overview of current state-of-the-art and/or latest research results in the relevant areas. National and regional differences or specific achievements are mentioned, where available. Discrepancies between vision and reality are not highlighted here, but serve as a basis for the issues identified in the subsequent section.

3.1.2.1 Intermodality – integration of modes

Intermodality is the all-important word for passenger transport systems. Intermodality stands for multimodal transport, incorporating completely linked and adjusted systems with fully integrated operation and management, availability of information anytime, anywhere for operators and travellers, and much more. As intermodality covers a very large area, related contents are only briefly mentioned in this section, but are treated more fully in accompanying chapters.

Major parts of the required physical infrastructure for the envisioned intermodal transport system are already in place and available. In some conurbations mass transit modes already work in an integrated manner within a common public shell to market their services, yet the ties between urban mass transit and long distance services are still loose. Integrated tickets for door-to-door journeys are not available. Different agglomeration areas have non-compatible ticketing systems. National borders limit seat reservations. Schedules are co-ordinated within individual modes and subsystems, not between different modes or operators. Information on schedules and actual performance is not exchanged between operators, so different operators cannot adjust their services in response to lateness of a connected service.

Passengers have to approach providers individually to obtain information, service and tickets from them. Quite often neither tickets, route nor schedule information are available from mass transit operators at the destination point of a journey. Value added services like baggage handling are linked to individual modes or not integrated at all into the transport network.

Figure 3 shows the different demands of today's transport systems, lacking intermodal connections for an overall integrated service.

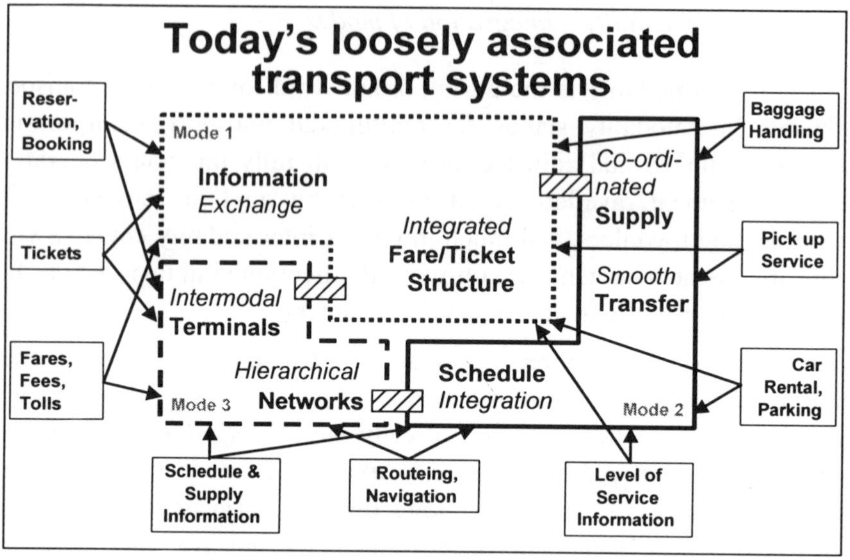

Figure 3: Organisation of today's transport systems

For the time being, the best practice of operators is to provide static schedule information on their transport service. That data sometimes is available across Europe by means of the Internet. However, quite often the data lacks reliability, and short-term changes of schedules are not available. Many operators are still reluctant to even inform travellers about lateness or unavailability of services within their own systems, though real-time information on the actual performance of individual services is increasingly being deployed in some urban systems.

The CIVITAS initiative funded by DG TREN aims for 'cleaner and better transport in cities', based on eight CIVITAS Policy Strategies that will be applied to the sites involved. Approaches include new demand management strategies, innovative logistics services and integration of transport management systems. The CIVITAS I initiative covers 19 European cities, with a further 17 cities in CIVITAS II.

3.1.2.2 Passenger transport systems and operation

The Public Transport Operations (PTO) area addresses mostly urban and peri-urban transport in both road and rail modes (including the mixed mode of trams), and it is thus related to the network and traffic management and rail areas, as well as to other traveller intermodality areas such as integrated payment.

Travel information services are discussed in section 3.2. In many current projects emphasis is placed on overcoming intermodal and/or cross-border limitations, aiming for multimedia, multimodal, door-to-door information for travellers.

Public transport priority

Demonstration of Public Transport Priority integrated with an urban traffic control system has taken place in a number of 5th FP TAP projects. An example is the TABASCO project, which also included the linking of urban and inter-urban telematics systems, stimulating modal change, working with public/private partnerships and advancing system architectures. The BALANCE adaptive signal control method developed within TABASCO showed encouraging results including reduced journey times for bus and tram services, reductions in delays and queue lengths as well as reduced emissions while safeguarding network conditions for other road users. Cost/benefit analyses forecast redemption of costs for modifications within a few years.

A project supported under the 5th FP was PRISCILLA that focused on dissemination of a best practice guide for bus priority in wide areas.

PRISCILLA aimed to demonstrate and facilitate the rapid take-up of experiences gained in previous RTD projects related to bus priority systems in small networks, enlarging on them to show the effects and impacts in wide networks. This included a state-of-the-art review of bus priority systems, implementation and demonstration of wide area bus priority via field trails, evaluation and assessment of user acceptance, impact and cost/benefits and finally dissemination of a best practice guide for bus priority action.

The outcome to facilitate the adoption of bus priority systems for wide areas has been satisfactory, but scope for further research was identified in the application and evaluation of different strategies under more varied traffic and bus network conditions, different traffic signal plans, bus priority parameters, bus detection systems, etc.

Priority of PT systems is now being widely implemented in a number of cities throughout Europe. However, optimisation strategies over large networks need to be further improved before best practice can be established. An interesting approach of on-demand PT priority has been deployed in Zurich, Switzerland. Here priority for PT is not automatic, but will only be given if the approaching vehicle is delayed and demands priority. This prevents giving unnecessary disadvantage (increased waiting times) to individual modes (e.g. cars, bikes, pedestrians) in situations where priority is not needed.

Vehicle Scheduling and Control Systems

Today there are many Automatic Vehicle Management (AVM) or Vehicle Scheduling and Control Systems (VSCS) used in Europe and overseas. A prototype dynamic bus scheduling and remote maintenance monitoring system funded under the 4th FP was demonstrated in Valencia (Spain) under the AUSIAS project. The dynamic bus scheduling application had two parts: off-line scheduling of vehicles and crews, and online re-scheduling in response to incidents, with software to recognise and manage 'regular' incidents.

Technical validation of the integrated fleet management system was performed successfully. User acceptance of the off-line application to schedule vehicles and crews was high and the time to create a bus and crew schedule was reduced by one third. The dynamic scheduling application could not achieve full assessment due to an insufficient incident database, but it was concluded that there was a good potential for development based on the initial results.

Project BERTA developed a basic tool for maintenance of all ITS data of bus, tram and metro in the city of Berlin. BERTA covers functions ranging from drawing timetables, monitoring actual operations, providing real-time information to passengers, to supervising safety and security in the stations. One tool (RBL) controls, manages and optimises bus and tram operation, the other (LISI) is an integrated safety control and management system for operation of metro systems. All systems work together online and are fully integrated.

Improved interoperability

A research project addressing data exchange for better public transport operation was TRIDENT funded under the 5th FP. The goal of the project was to support multimodal travel ITS by establishing the common and reusable mechanisms that are required for sharing and exchanging data between transport operators (content owners) of different modes. It also investigated and proposed solutions for well-known organisational and strategic issues hampering travel intermodality.

This led to proposals for new standards, as well as development of specifications and software modules which enable the sharing and exchange of real-time multimodal traffic and traveller information through the whole Traffic and Traveller Information (TTI) content chain. To achieve this goal, two different paths were selected – a 'messaging approach' (EDI, DATEX) and a more modern object-oriented technologies approach. The two sets of specifications have been implemented in test sites in Flanders, Paris, West Yorkshire and Rome. Applications have been successfully tested, trialled and modified and have continued operation after the end of the project, in some cases already extending to other areas and transport modes. Specifications have also been submitted to CEN TC 278 working groups. (CEN (Comité Européen de Normalisation) 278 fosters European standardisation between systems concerning road traffic and transport telematics.) For further development a post-project platform TriEx has been founded.

Improved interoperability of maintenance

A completely different but indispensable aspect is maintenance of passenger transport vehicles.

In the 4th FP project AUSIAS, an automated fleet maintenance tool was developed for monitoring bus engine parameters in real time. Sensors are constantly checking the function of the engine; when a deviation of a parameter is detected, an alarm is sent to a co-ordination centre including the information on the vehicle's parameters. The incident is analysed with the help of additional information on this vehicle gathered previously (e.g. problem history, repair history). When the diagnosis is complete, possible faults and recommended actions to correct them are presented to the driver.

The 5th FP project EUROMAIN aimed to define, implement and validate a complete maintenance support system for railways. The project will allow the monitoring and diagnosis of complex equipment aboard trains as well as inside fixed plant, creating comprehensive management tools for maintenance. Outcomes include proposals for new standards as well as assuring cross-border interoperability for systems and operators. This includes standardisation of formats and models for data exchange, remote monitoring and diagnostics, rules for technical documents, interfaces and specifications for maintenance and configuration management, as well as optimisation methodologies.

As the actual state of vehicles will be provided through real-time information, this will remove the need for fixed schedules for maintenance, reducing maintenance down time and costs. Based on a common cross-border approach, co-operation and document exchange as well as experience achieved, this may contribute to the establishment of a trans-European railway network.

3.1.2.3 Payment systems

Smart cards are now a common reality of the daily life of many citizens. They are used extensively by the banking sector as debit cards, credit cards, and electronic purses. They are also used to identify subscribers to services or associations; they are part of our mobile phones, and now in some countries they are used in the health sector to identify a person and contain some useful information on his/her medical file. Nevertheless widespread usage of multi-application smart cards, especially for passenger transport, still remains a future vision in most parts of Europe.

Applications in the transport sector in Europe are growing, and focus on the areas of payment and ticketing. A device used for payment will handle information concerning the identification of the paying person and/or the account to be debited. For ticketing, the device must also include the identification of the travel paid as evidence of the contract between the transport service provider and its client.

The integration of multiple applications on a single card or other mobile unit (e.g. mobile phone) have been pursued so as to avoid situations where people need to carry numerous smart cards, each devoted to a special service. This has led to the emergence of the concept of the 'city card' where a citizen can access public transport, canteens, libraries, swimming pools, and other types of public services within an area using just one card.

The European project CALYPSO starts from the ticketing/payment function in public transport. The smart card can be inserted in a contactless folder called the Maxipass, which provides information on public transport travel through a small screen (time of departure of the next train, events on the line, best itinerary, etc.). A simpler version of the CALYPSO device is just a contact and contactless smart card called Minipass without any folder, essentially for payment-oriented applications. The chip contains an electronic purse and the contactless antenna allows debiting of a central or an onboard account.

TELEPAY uses a different approach to innovative payment systems for transport services. As a smart card is relatively expensive, contactless smart card based ticketing systems are only cost effective for multi-ride tickets and season tickets for regular customers. Therefore the TELEPAY concept is based on a combination of widespread mobile phone technologies (GSM/SMS, GSM/WAP, GPRS, Bluetooth) applied to a new application. The project aims to trial this innovative approach for the payment of public transport, parking or the toll on motorways.

The realisation of the system encompassed the adoption of wireless devices as a payment means for transport services, establishment of technical, legal and commercial feasibility of purchase and payment, and implementation of the system in four test sites across Europe, as well as assessment of systems including identification of future tasks. Systems were broadly accepted, while at one test site transport authorities even continued the service on a commercial base. Other interesting output was the comparison of different implementation approaches through the test sites, covering technologies, but also public organisation, transport services and service organisations.

The TRIANGLE project validates the concept of a simple, workable and manageable interoperable solution for door-to-door travel (combining payment, ticketing and access to services) using existing technology and/or installed infrastructure (e.g. CEPS, EMV, CALYPSO, IOPTA) and is compatible with ongoing programmes. The solution is based on smart card technologies that allow various applications, such as transport tickets, electronic purse and services to be hosted on a single chip. The objective is to prove the viability of truly multi-applicable smart card schemes integrating features such as cross-border interoperability (e-purse and ticketing) or advanced booking for long distance trains. The concept and the performance of existing technologies should be proven in two steps, a laboratory test as well as use in the cities of Brussels, London and Paris and the international rail service linking them.

Standardisation for smart cards is already beginning to be addressed to allow interoperability between more and more applications in the field of PT services. Interoperability between cities is another issue, which might come out of the standardisation activity but does not look essential today.

To the east in South Korea, Hong Kong or Singapore deployment of smart cards is more advanced. Tag-on/tag-off on bus services with variable fares and transfer rebates have worked very well in Singapore. Boarding times on Hong Kong buses are very low. In South Korea the means of paying for public transport has changed completely across the country, as the use of smart cards has become part of everyday life. Smart cards for surface transport fare payment are also in use in Asia (e.g. Malaysia, Bangkok), and increasingly in Australian cities as well.

3.1.2.4 Demand responsive transport services

This domain covers mobility services that are made available to enhance citizens' possibilities of accessing flexible and collective transport services. Demand Responsive Transport (DRT) serves city core areas but also rural areas. Today services are predominantly specialised on areas and periods of time that traditional services have difficulty in covering and user groups that have difficulties in using traditional public transport. Recently other approaches have also focused on offering an alternative to regular PT services through on-demand door-to-door transport using minibus vehicles.

The research in this area has addressed the provision of flexible transport services including details of the technologies, user acceptance, institutional and organisational aspects, as well as the business case and best practises in managing and supporting these services with new applications, systems and communications. The major part of the work was carried out in projects to develop and validate Travel Dispatch Centre (TDC) concepts and technologies to manage multi-operational and multimodal transport services (i.e. taxis, taxibuses, minibuses, vehicles with equipment allowing access for people with wheelchairs) with the integrated enhanced booking, reservation, route optimisation and vehicle dispatch functionalities.

Examples of successful DRT projects are 4th FP SAMPO and SAMPLUS which have demonstrated and validated the DRT systems on test sites in Belgium, Finland, Italy and Sweden. The operational environments covered rural and urban areas. Institutional and organisational issues have shown to be crucial for successful implementations. However, the rapid technological developments achieved in the projects have created new supporting tools for implementation of DRT services, including transport services for the disabled and elderly, as well as for extension, feeder and/or replacement services to scheduled collective transport for the general public.

The main results from these projects are the enhanced functionalities for existing and newly created Travel Dispatch Centres (TDC) that establish an operational environment where local authorities have the means to provide, manage, control and plan transport services for citizens, with less financial support and with better quality of service. These functionalities include a superior reservation system, using geographic information systems (GIS), managing the customer databases and using smart cards to pay for the services and as an ID for automatic reservation of the return trip back home. On the hardware side this resulted in increased processing capacity, improved reliability of the system, improved management of the operations and enhanced automatic vehicle location (AVL) technologies. The economic viability of DRT services is seldom based on the fare box revenues. The financial justification for DRT comes from the fact that instead of an annual increase of 15 % in costs to provide special transport, services are enjoying an annual 2-3 % decrease, including the costs of the TDC and its equipment. After the projects finished the demonstrations were turned into commercial DRT services. The patronage has grown rapidly reaching a level four to ten times higher than when the services were introduced.

Though the suppliers developed their products in collaboration with each other using the same global specifications for functional, informational and physical architecture, different operational and technological environments of the test sites led to quite different approaches. This has enhanced their competitiveness on the global market place, especially when the requirements of new customers may significantly differ from those of European customers. Projects have also established various contacts with potential customers in most European countries as well as in the US and the Asia-Pacific Region.

The trial project FAMS is aiming at scaling up the technologies, service and business models currently adopted in DRT. FAMS supports the evolution from single DRT applications towards the concept of a flexible agency for Collective Demand Responsive Mobility Services, which is crucial for further expansion in wider areas.

E-business/e-work collaboration along with team-working tools and methods should create a Flexible Mobility Agency that co-ordinates the different operators and organisations as well as enhances the accessibility to mobility services for users. Information exchange is based on standard Internet and web technologies to provide flexible and dispersed interfaces for customer services. For DRT operators, Internet-based services enable concerted and flexible co-ordination of services like vehicle or staff management, resource planning, etc. The main vision supported in FAMS is that all actors of the DRT service chain, both the different transport operators and the different user groups, constitute a virtual community.

3.1.2.5 Intelligent Grouping Transportation (IGT)

Another interesting approach, deriving from a Greater London case study, is Intelligent Grouping Transportation (IGT). Instead of concentrating on times and areas of low demand, this approach focuses on offering an alternative to nowadays established means of motorised urban transport such as individual cars, taxis or traditional public transport.

Assuming an adequate number of vehicles, the overall transportation efficiency arises from the method of grouping passengers with compatible itineraries into the same minibus vehicles. A computerised system organises the grouping so that the vehicle can transport all passengers on-board from door-to-door, thus being independent of additional infrastructure like bus stops or metro stations. Payment is realised through automatic fare charging.

The IGT system is based on four elements – transport of passengers through 'Taxibus' vehicles, booking of trips (on-demand/pre-order/regular order booking) via mobile phones and networks to transmit the data, GPS in-car satellite navigation to provide exact real-time location for guidance of taxibuses and computer systems to manage the taxibus fleet.

3.1.3 Issues

This section highlights the research, administrative or organisational issues required to move from the current state of research and technology towards the visionary situations mentioned in the first section. It is evident that results of current research projects are not yet the common state of technology in Europe, so a lot of effort is required to disseminate these results to transport operators, political and administrative bodies, industry and users.

3.1.3.1 Intermodality issues

The following issues could be addressed or at least largely improved by means of telematics. However, it is important to remember that user needs should always have priority over purely technical advances. These issues are not specific to intermodality, but are also addressed in other parts of this section, or of the book:

- Traveller information (also see section on 'Information Services'):
  - Pre-trip information: comprehensive information at bus stops/stations, providing information on next vehicles, lateness, loading, special services, incidents in the network;
  - On-trip information: in-vehicle information displays showing whole/section of a network, rather than just the next stop; advice on best onward connections.
- Integrated booking/ticketing (also see section on 'Payment systems');
- Electronic payment (also see section on 'Payment systems');
- Real-time vehicle monitoring and schedule information (also see section 3.1.3.2 on PT operation or section 5.3 on 'Road and traffic monitoring').

Major challenges, that have to be resolved before telematics application can gain further ground in intermodal transport systems include:

- Clear concept for the organisation of the transport market; liberalisation where market forces and competition improve efficiency and regulation of the market, where necessary, for public interests;
- Dedication for the implementation of the concept above and clearly organised administrative responsibility for operation of transport services;

- Standardisation of ITS architecture;
- Unsettled issues concerning the availability and ownership of data as well as the exchange of information between different service providers;
- There is a need for further research concerning operator/operator and operator/authority co-operation;
- Further studies on unified tariffs and presentations – especially for intermodality within large networks with several competing operators and different modes.

A further aspect concerns the effects of variations from the scheduled arrival and departure times. How easily can connecting services react to lateness? How much time needs to be allowed or wasted at interchanges? Would there be a benefit to travellers if arrival and departure schedules came enhanced with standard deviation measures?

Another issue is the critical assessment of a typical user's capability of actually understanding and utilising advanced intermodal passenger transport service information. The user needs to access and assess large amounts of information both physically and mentally and is required to choose from an ever increasing set of options. While moving, orientation in vehicles and interchanges is crucial to seamless travel. Information overflow is a regular feature of the information technology society and a competitive market will neglect any concepts or systems that are too complex to understand. The same applies for wide area networks that are operated by different service providers. Common design and functionality should be provided for the whole system to ensure seamless access and utilisation. Unfortunately, the market for passenger transport services is not competitive at large, bearing the risk of wasting public subsidies for services that are not utilised due to their overwhelming complexity. These issues should be carefully studied to properly allocate any funding.

3.1.3.2 Passenger transport systems and operation issues

In the field of Public Transport Priority, identification of best strategies for overall network optimisation will need further analysis and trials as part of some take-up actions.

Integration between Public Transport Operations (PTO) applications and intelligent vehicle applications are issues for further research and development ('Intelligent bus', 'Intelligent train', etc.).

Especially in wide area transport networks there is a requirement for advanced monitoring and control systems that are able to operate in a decentralised way, offering different levels of information and functionalities for operators, along with concepts integrating systems to provide truly seamless services. This is to be accompanied by a comprehensive data warehouse concept that collects, holds, manages and distributes all data for relevant applications.

Therefore the most challenging aspect is the improvement, integration and processing of information generated by monitoring, management and control systems. Real-time data will be prerequisite for flexible matching between current travel demand and services offered. This is to be enhanced through tools forecasting near future conditions of the system, especially valuable in case of deviations. To offer advanced services, there needs to be a better understanding of travel patterns, obtained through surveys but moreover via new tools of data collection with input from booking and route planning systems, smart cards, event or complaint management, etc.

PTO applications will have to be integrated with ubiquitous traveller information services that respond best to the real needs of users throughout their intermodal journeys. Fleet management, personnel rostering or Vehicle Scheduling and Control Systems (VSCS) would also need further integration with both integrated payment/demand management applications, and some aspects of demand responsive transport systems that best meet the needs of the users and service possibilities of the operators in certain networks under certain conditions (e.g. integration of taxis and public transport services in suburban areas/off-peak periods).

Also in urban areas, artificial intelligence and decision-support techniques could be usefully applied in support of automated vehicle location. Buses do not compete well against personal cars because of lack of speed and comfort. More information should be given to passengers about available seating, providing more comfort through a better distribution of passengers among buses, as well as exact departure and travel times. Guiding systems must also be developed to support bus movements.

In conjunction with the above elements of public transport services requiring new telematics applications, the following aspects need to be considered:

- Conditions for acceptability by operators;
- Conditions for acceptability by passengers;
- Ergonomic aspects;
- Public-private partnership;
- Legal support;
- Technical standards;
- Design and operation standards;
- Economic viability.

Further technical problems which need to be addressed to improve public transport operation are discussed below.

- *Real-time in-vehicle information on network performance:*

  There should be more information available to passengers (acting on real-time data) than just the name of the next stop of the vehicle. This could be information on lateness of connecting services, on incidents in the network or special services to inquire about re-routeing whilst on the trip.

- *Real-time evaluation of the load status of vehicles in terms of number of passengers:*

  Different possibilities exist, such as transmission of information measured on the suspension of the vehicle, counting of passengers (entrance and exit), in-bus video safety systems with image analysis and evaluation of number of passengers, or weight measure of the bus from the road, knowing the empty weight of that type of bus. Information is then transmitted giving time to arrival and load status (seats available, standing places available, full).

– *Software for demand responsive semi-public transport:*

  This software must be multi-operator and multi-vehicle, reporting to the transport authority. It must be based on a public-private partnership (PPP) principle, and linked to an automated transmission to the vehicle, with acknowledgement returned to the centre.

– *Intelligent vehicle maintenance:*

  Advanced in-vehicle sensors and recorded operational data should be connected to a vehicle management database to create an intelligent vehicle maintenance system. This enables the change from fixed intervals to more incident-orientated, integrated and dynamic maintenance, thereby increasing the fleet's capacity. More sophisticated systems require well-trained operators and staff, including adaptation of working procedures between linked companies. Education of staff has to foster an understanding of purposes, strategies and potential impacts of the measures, possibly by means of self-training and simulation tools. In case of failure, systems should be well supported by backup systems and/or substitute procedures.

3.1.3.3 Payment systems

For smart cards, both contact and contactless, the need is now for standardisation at application level and for the exchange of data. Work has started, but needs to be finalised in terms of technology as well as for the implementation of interoperable services, which might lead to necessary contractual agreements between service providers. Smart cards have proven to be an active part of our daily life. The TAP programme has facilitated their application to public transport and other city business in some European cities. In this way, it has also helped active strategies to improve the use of public transport in our congested cities, fully in line with policies developed all over Europe. It has provided European industry with the new elements necessary to maintain a high position in this domain of high technology. Standardisation and interoperability of electronic payment systems at national and European levels still requires major efforts. The common currency should help to facilitate standardisation throughout Europe.

However, development of advanced payment systems is not just concentrated on Europe. Overseas, such as in Asia, utilisation of smart cards is enjoying great popularity as well as a high status of development. This has implications for European equipment and chip manufacturers, since those large markets are also setting specifications and performance levels.

3.1.3.4 Demand responsive transport services

Missing legislation is probably the biggest obstacle in most European countries. Currently, legislation strictly divides services into route-oriented public transport services (fixed stops, fixed schedules, fixed fares) and taxi services. Some countries do not even use shared taxi services, picking up additional passengers along the way. Subsidised demand responsive transport is cutting severely into the taxi market, raising questions of governmental influence on an ostensibly free market.

Just as much as DRT overlays with existing taxi services, integration into regular public transport also raises many questions. This ranges from different (usually smaller) vehicles required for DRT, subcontracting public transport or taxi operators for the actual driving, to integrated fare systems for both DRT and regular PT or carefully adjusted schedules allowing comfortable onward connections. Different levels of public subsidy for taxi, regular public transport, demand responsive transport and special needs services must be considered and integrated into an acceptable fare structure that enables adequate pricing as well as potential for flexible strategies, such as loyalty schemes or temporary special offers for the promotion of special lines. The challenge is to generate a balanced system of modes serving different needs. In wide area networks especially, this will involve different service providers that have to be covered under one shell to ensure seamless operation and service by means of advanced B2B and B2C applications. Best practice guidelines for phasing and operation and also business models including cost-benefit analysis are key for wider deployment.

Building an actual market for DRT services is yet another challenge. Public transport users in general are not used to flexible transport supply, requiring advanced booking of a trip. While this is quite common in air transport and long distance rail service, passengers need to get used to the idea for local transport. Specific marketing efforts for different user segments are required and supported by dedicated surveys. Passengers need to be reassured that the service and the fares are reliable. Specific training may be required for advanced systems with automatic or semi-automatic localisation and booking procedures.

3.1.4 Future opportunities

The previous section sought to present the most important issues required to bring passenger transport services into life – regardless of ITS contribution. This section recognises the ITS research and development activities already taken up and presents the most important and interesting opportunities for ITS in development of passenger transport services.

3.1.4.1 Intermodality – integration of modes

Major obstacles on the road to a truly intermodal transport system stem from incompatible institutional set-ups that will not be resolved with technical means, but by political will. European and national legislation need to define the framework for international, multimodal co-operation of transport service providers. Sustainable mobility and public welfare require a careful balance of market forces and administrative regulation. A network of excellence made up of high-ranking political and transport operator representatives (no research bodies, no consultancies) should prepare the grounds for adequate legislative actions.

Competing providers on an open market are not necessarily interested in co-operating with others or providing detailed information on their service performance. Liberalisation and deregulation do not automatically lead to one simple, efficient, integrated multimodal transport system and operators need to earn a reward or gain a profit for thoroughly co-operating in an intermodal system.

An indispensable prerequisite for the integration of transport modes is the abundant sharing of operational data – offline planning data and real-time information on actual performance between different modes and operators. This touches questions about organisation of transport markets, co-operation and competition of operators, value and ownership of data, but also technical definitions and common standards for the exchange of data. National standardisation bodies should be invited to provide valuable input to a network of excellence, but must avoid generating several, incompatible solutions for the same task.

Electronic planning, operation and service tools have become quite common throughout Europe. As yet these programs usually address one specific task each, such as exchange of data between planning, passenger information, ticketing, operation and fleet management. This leads to inconsistency of data and little flexibility in case of incidents. Standardisation of transport operation data provides the opportunity to build on system solutions that not only integrate all the tasks mentioned above but also have the flexibility and power to handle a large number of modes in an integrated network. The integration of public transport operation tools should initiate development in the area, while minimising the risk of incompatible, proprietary solutions.

As a spin-off from shared real-time operational data, multimodal passenger information could be vastly improved. Besides all sorts of personal travel services, public on-trip information should be strengthened. In-vehicle information displays, PDAs, mobile phones or interactive kiosks should provide information on onward connections, delays and incidents as well as general infotainment contents.

From the customer's point of view an integrated fare structure and through ticketing are most important elements of an integrated multimodal transport system. From the operator's side this requires sophisticated tools to monitor passenger flows and share revenues. A common fare deprives operators of an opportunity to distinguish themselves from competitors. The general public still has a vested interest in socially acceptable transport fees and needs to address equity or subsidy issues.

A broad, scientific analysis of fare structures applied throughout Europe should form the basis for a generally applicable tariff model, even for large multimodal transport systems, providing a balance of user interests, operator needs and public concerns. This could be an area where national projects should be supported.

3.1.4.2 Passenger transport systems and operation

Even with promising results in the area of public transport priority and vehicle scheduling and control systems achieved in previous research and demonstration projects, there is room for further improvement and integration of systems and operators. In the field of public transport priority in urban traffic control, identification of best strategies for overall network optimisation will need further analysis and trials as part of some take-up actions.

Among the many tools convenient, but yet unavailable for operators, is a comprehensive, automated economic assessment and optimisation algorithm to develop complex PT networks and schedules. Automated production of route databases, timetable databases and vehicle and driver schedules would be essential sub-tasks to that system. Potential transport markets would be identified out of GIS based structural data and taken as a basis for automated line and network definition.

Dispatchers' real-time decision-support tools for re-scheduling and re-routeing due to incidents build on that database and allow more flexible operation. All data should be accessible immediately for subsequent operating systems, handling fleet management and maintenance as well as for information systems providing operators, staff and travellers with the latest information. Actual travel demand, as derived from ticketing and loading information, needs to be fed back to the system immediately so that overall performance of a PT network could be iteratively improved.

Public authorities require access to the system and data to control performance and quality of operators and services, enforce public interests and standards and also monitor and prosecute any licence violations by competing operators. A network of excellence could work on improving the different aspects of the public transport planning and operation software tools and ensure that every module and sub-task fits into a common system architecture and data structure.

Besides improving the operation of existing transport modes the market potential of several new systems needs to be explored. Guided buses, tram-trains, innovative/automated passenger transport systems or people-movers provide special solutions, sometimes with explicit advantages such as fully automated 24 hours service, low environmental impacts, less demand of space or costs for infrastructure.

3.1.4.3 Payment systems

European citizens increasingly expect to pay by electronic means for goods and services – even so in transport. However, there are some distinctive requirements to that field that must be solved, before any system can be applied successfully in a pay-per-ride manner:

- The passenger needs some sort of proof that he actually paid his fare;
- Boarding must not be delayed by the payment/enforcement procedure;
- The payment system must comply with the local fare structure, e.g. in most networks the distance or number of stops influence the fare;
- Frequent users expect weekly or monthly travel passes with limited costs or other refunds and special offers;
- Through ticketing in a multimodal, multi-operator environment is an essential;

- Personal data protection prevents operators from storing customer data that could be used for tracking a traveller's movements;
- The system needs to handle frequent users and occasional visitors alike;
- Conventional payment and ticketing will need to be supported in parallel for a long time.

Current development of passive, contactless smart cards proves promising in meeting the requirements, but further work on integration of numerous functions and standardisation is required. In the long run it can be expected that (general) payment systems, personal travel assistance and mobile communication devices will merge into single units. While this development is pushed by the banking industry on the one hand and the electronic device manufactures on the other, attention should be given to the appropriate consideration of specific transport requirements. An integrated project of major European public transport operators, cities, industry and banks should create the impetus for widespread introduction of contactless, electronic payment in public transport. As issues of commerce and transport are not limited to Europe, experience and expert-knowledge at the international level should also be involved.

3.1.4.4 Demand responsive transport services

Demand responsive transport services are a specific and very promising form of passenger transport operation – with greatest potential in low demand situations like rural areas or urban night services, but also serving as a potential alternative for traditional transport modes, especially where not yet deployed.

So far it must be clearly stated that technology issues only provide one, albeit important, facet to the development of integrated DRT services. Especially in wide area networks with possibly different service providers, it is acknowledged that relevant obstacles to large-scale implementation can be identified in different areas, like the legislative framework for DRT missing in most European countries or the establishment of the level of community welfare required for lasting public subsidy of such services. Building an actual market for DRT services is another challenge - many people are not used to the idea of flexible transport supply.

Smart cards could be incorporated with DRT services in a number of ways. They could help with identification and localisation of customers, act as electronic tickets and support electronic payment of fares. Registration of passengers with the help of smart cards could also provide proof of actual transport performance provided by specific operators and facilitate clearing procedures.

A major issue in the field of technology is the standardisation of necessary equipment, building an open market with competing enterprises in the industry. Standardisation of interfaces and some of the components will raise competition and provide better, cheaper solutions for widespread application of an open market.

This ranges from cheap, powerful data communication between vehicles and control centres to fully automated booking and dispatching systems. There should be specific scientific research on dispatching tools for (wide area) demand responsive transport, developing the heuristics necessary for dynamic re-routeing of vehicles and precise forecasts of pick-up and arrival times. In this context, institutional services between public transport authorities and private operators have to be examined. HMI issues arise both on the customer and on the driver side. Existing technologies like GPS localisation in GIS systems need to be better adapted to the specific needs of DRT operators.

3.1.4.5 Non-technical issues

In addition to the technical research on intelligent transport systems, it would be worthwhile to initiate some more empirical research on the social aspects and contributions of ITS. Equity and welfare effects should be studied to properly allocate public funds. Most of the technologies discussed above require large investments from operators or from public funds, while the benefits usually remain with the users. A thorough discussion is required as to whether the benefits to the general public justify investments from public funds or if the actual users should finance the improved performance of systems through adequate user fees. Cost-benefit studies on ITS investments seem to be rather sparse. Analytical instruments to measure changes in user behaviour, improvements in system performance or social benefits initiated from ITS measures are lacking. Empirical surveys could help to provide better understanding of current and future travel demand, including from occasional users (e.g. car drivers) as well as referring to future aspects such as changes in society or environmental concerns (White Paper – European transport policy for 2010, Kyoto Protocol to the United Nations Framework Convention on Climate Change).

3.1.4.6 Conclusions

European research from the Framework Programmes on RTD has already provided a wealth of results, but these are not yet commonly acknowledged. It will take a lot of effort to disseminate these results throughout Europe, forming the basis for the next level of research. Efficiency of funding in terms of public benefit would probably be higher through promoting the technologies already developed than through supporting further cutting-edge research. Industry needs to get involved in feasibility studies and larger field trials are required to get public transport authorities and transport operators alike interested in investing into advanced and more user-friendly systems. Support for larger field trials, demonstration projects and elaboration of realisable business cases could be the proper tool in an environment proving to be generally conservative towards rapid changes and rightfully reluctant to invest public money in unproven benefits. If new systems are not likely to be tested or deployed in Europe, deployment abroad should also be fostered to create examples for real operable systems and environments. A well-funded and active information campaign, promoting the solutions already proven in practice, could be the most beneficial measure to increase the application of ITS in public transport services.

3.2 Information Services

3.2.1 Background

Traffic and traveller information services (TTI) have been one of the fastest and, for the general public, most visible areas of growth in transport telematics in recent years. In addition to the traditional sources of travel information, a plethora of websites offer journey support: from planning and routeing, to reservations and ticketing. During a journey, automated navigators help find the way, while real time updates on travel conditions are provided by infrastructure and transport operators. The information itself is delivered via numerous channels: onboard units, the Internet, mobile devices, etc as well as public message systems. So what is there left to do? Is there still need for R&D at the European level?

Despite the progress made, there remain some significant shortcomings in TTI services – regarding the information itself and the *way* it is made available. In a context where 30 % of transport users say that they would have chosen a more convenient way of travelling if they had had better information about their options (ERTICO booklet 'The Vision', 2002), there is clearly room for improvement. Among the areas where further progress needs to be made are:

(a) Improvements in the *accuracy and reliability* of information, so travellers can be fully confident in its use;

(b) Better information on *multimodal options*, so travellers are encouraged to choose public transport and non-road modes;

(c) More timely information on *delays, incidents, and service disruptions*, to give travellers 'early warning' and information on alternatives in order to facilitate changes of plan;

(d) Improved *delivery of information,* to make messages easier to understand and available via multiple channels;

(e) Continuity of services *across national borders*.

An attempt is made in this section to identify the obstacles - technological, organisational, legal and other - currently hindering the achievement of these objectives. Suggestions are then made relating to the research and other actions which could help to overcome them. A further aspect to be explored is the possible strategic use of transport information as a *tool for traffic management*.

Finally, a series of recommendations are made consisting of initiatives which it is felt should be taken at European level to favour the achievement of the 'vision' described below.

3.2.2 Vision

The three applications described in this section are intended give an idea of the practical implications of traffic and traveller information services which could be used in the future by travellers in Europe.

Giles is a French businessman who lives in the country, but has an office in Lyon. He normally goes to work by car and subscribes to an information service which sends him a warning when there are problems affecting his usual route. After heavy snowfall on Sunday night, he receives an SMS early on Monday morning to warn him of a serious accident and tailback. He immediately consults his service provider's website and is presented with the details of two possible alternatives for his journey. Examining the trip times and cost, he opts for the fastest solution which consists of the train from a nearby station, then a bus or metro for the last part of the trip. When on the train and entering the city, he checks the arrival times of his connecting services using his laptop and sees that it is best to take the metro as the bus is running behind schedule. He finally arrives in the office 20 minutes later than usual, but in time for a planned meeting with a customer. He discovers at lunchtime that a colleague who came in to work by car from the same village reached the office three hours late.

Jaak and his family from Stockholm have decided to go to Prague for a long weekend. A month beforehand, they explore the Internet and find a site which gives useful suggestions for flights and accommodation. Since they have used this particular service before and qualify for the loyalty scheme, they receive special rates for their hotel booking and are offered discounts in certain listed restaurants. The site also gives them ideas for possible trips when they get to Prague. They decide to buy tickets for an exhibition as the site tells them it is already heavily booked. When they arrive, they find their smart phone very useful on trips around the city as it helps them find their way by showing images of distinctive landmarks. Although the navigation service is run by the local transport agency, they can receive the directions in English, which makes it easy to use public transport. They are also pleased to be able to use the same smart cards for their fares and meals. The only problem they encounter is with the prepaid exhibition tickets which they find are not valid on the day they want to visit. When they get home they are able to follow this up, making a claim for reimbursement via the same website.

Steen is a driver for a Danish company which manufactures furniture. He has been given the task of delivering a consignment to a warehouse in the outskirts of Barcelona. The company's trucks are equipped with onboard units which support fleet management, but also incorporate a navigation and information service. When crossing Germany he finds this very useful for re-planning his route to avoid busy sections of motorway. As he enters Italy, it is also able to warn him of restrictions in force on Sundays on heavy traffic in Italy and to help him find places to eat and stay overnight. When he arrives in Barcelona he requests authorisation to enter the restricted traffic zone and is able to book a convenient unloading bay. To guide him there he receives spoken directions in Danish, which means he can focus on the traffic without distraction.

3.2.3 State-of-the-art

3.2.3.1 The TTI market

In less than a decade, the market for TTI services has grown from a non profitable niche market to a substantial business. Many city, regional and national platforms are now in operation, and an increasing number of services are provided by public private partnerships (PPPs). Difficulty is still encountered however in establishing commercially sustainable services.

The main types of player involved are:

(a) Transport and infrastructure operators, who collect and deliver information specific to their own services/operations;

(b) Public authorities at the city/regional/national level who provide services catering for the needs of travellers in their area (i.e. covering all modes within a given geographical zone);

(c) Private service providers, who address the needs of specific market segments.

A comprehensive survey of the TTI market in Europe was undertaken by the ATLANTIC project. The results, published in 2002, suggest that the tendency is to follow the sequence indicated above: beginning with type (a), then moving to (b) and finally including type (c). A sign of maturity is the presence of different types of service provider and, in particular, the involvement of the private sector in data acquisition and processing as well as in its packaging, transmission and support services. Six different service models were identified, their features depending upon the stage of market maturity reached and the policy orientation in the country concerned. Two main clusters were distinguished. The former, more typical of Southern and Eastern Europe, has a predominance of public and single-mode services, while the latter, typical of Western and Northern Europe, is characterised by a rapidly growing number of private TTI services.

We now look at the situation in greater detail.

Most operators of public transport, motorways, roads, etc. now deliver information to the public on their own services or network. In some cases this is also shared between operators (e.g. transport services in the same city, concessionaires within a national motorway network). It is made available through various channels, but web-based services are dominant. The information itself is generally free of charge, although users pay for Internet connection time and personalised messages.

Among the many city transport information platforms in operation at the time of writing are StadtinfoKöln in Cologne, Trafikanten in Oslo, Trafikinfo in Copenhagen, 5T in Turin, and Transbasel in Basel. National platforms include Transport Direct in the United Kingdom,

YTV in Finland, OVR in the Netherlands, CCISS in Italy, CHAPS in the Czech Republic, and DGT in Spain. Such platforms collect transport data within a geographical area (in cities this is often multimodal) and make it available to the general public, usually free of charge.

There is enormous variability in the quality and availability of the data itself. Services offering high quality information and easily accessible appear to be used more intensively, but reliable statistics on actual use are seldom available. With the exception of special services offered to car drivers, evidence exists in only a few cases of willingness to pay for information (e.g. the 5T service in Turin reports that around 40,000 users a month send - and hence pay a small fee for - SMS requesting real-time information on bus and tram arrivals at city stops).

A look at such TTI services currently offered in Europe shows that:

- Very few are paid for directly by the final customers (either they are sponsored, or revenues are obtained from advertising or by selling information to other operators);
- At present, most TTI services operate within a single country (and more often a particular city or region);
- The majority of services cover a single mode; real 'multimodal' services are still rare;
- The transfer of information from transport operators to service providers is based on bilateral agreements; there is no standard framework.

Business models

The different types of service provider have very different aims and market approaches. While there are as yet no consolidated business models, two main categories can be distinguished. The first (prevailing) approach consists of TTI services offered to travellers free of charge:

- *Services provided by public agencies or transport operators*

  The agencies/operators are giving information on the transport services for which they are responsible. The data content is obtained from the control and management systems, and therefore with limited extra cost. Resources for providing the service are normally found within their budget, with the justification of 'marketing' the transport service or improving mobility. In all cases the cost of the service is in reality supported by taxpayers or by all users of the transport service. The information does not normally address specific market segments or personalised requirements.

– *Services available via radio broadcasts or Internet portals*

  Here too the user does not normally pay for the information itself. The broadcaster (or portal operator) obtains revenue from advertising, which is in turn dependent on patronage of the channel or website. Already common for the RDS/TMC service, this model is now being adopted in the world of the Internet. Due to the lack of TTI content at reasonable cost, it is normally limited to simple services. Users who require personalised data generally have to subscribe to a specific service.

In both cases the content is generally provided by public authorities and is limited in coverage, detail, and quality. In neither case can the model be said to constitute a real 'business chain'. The revenues are too small to generate sufficient value to create a market for information content.

The second approach consists of TTI services which generate revenues in the form of fees paid by users. In this case too, there are two main forms:

– *A 'vertical' structure,* i.e. where a single service operator covers the entire value chain: generating data content, providing equipment, operating the service, delivering information and customer care (e.g. Trafficmaster in the UK, TargaInfomobility in Italy). In general these services charge for the installation of an onboard unit plus an annual subscription fee. TTI is often part of a range of services for drivers including route guidance, live traffic information and tracking of stolen vehicles. Heavy investment is required to set up such services. This is justified when the information is valued by customers and other ways of obtaining it do not exist.

- *A 'chain' of operators with revenue sharing agreements*, which generally include the following elements:
  - Content provider (providing raw data about a transport fleet, a road network or part of it, or other data);
  - Content aggregator (who structures, stores, and sells the information about a given network);
  - Application provider (who develops and sells the service software and the platforms for service delivery);
  - Technology provider (responsible for the equipment used by the user – navigation/RDS/TMC receivers/personal navigation devices, etc.);
  - Service generator (service delivery/billing).

The role of content provider is normally played by transport authorities or operators. Content aggregators are often regional/national platforms, PPP or private service operators who establish agreements with the content providers. In some cases they also have their own content sources (e.g. floating cars). Application providers are usually software companies. Almost all of these operators act as both service generators and service deliverers, at least for a specific market segment.

At present, the services are generally delivered by one of the following:

(a) Content aggregators;

(b) Generic web portals;

(c) Broadcasters (for non-basic RDS/TMC);

or, more recently:

(d) Telephone carriers, offering travel information in combination with other location-based services, sometimes via multipurpose web services.

Revenue-sharing arrangements are set up between the service generator, content aggregators, content providers and – in some cases – application providers (technology providers usually obtain revenues directly from equipment sales.) When a fee is paid by the user for the TTI service, the mechanism is generally straightforward, but otherwise it can be rather complex and requires well-defined data exchange contracts.

In some cases, such as RDS/TMC, i.e. the travel news delivered by radio, no fee is charged for receiving TTI (at least for non encrypted services), on the basis that if broadcasters did not deliver quality services, the receivers would not be sold. This is leading to a solution where revenue transfers are made directly by the technology provider (in-vehicle equipment) to the service generator.

3.2.3.2 Market trends

Until recently the main commercial market in the TTI sector consisted of autonomous in-vehicle navigators. Today the market is far more complex and radical restructuring is occurring on the supply side (firms merging, changing roles, and new ventures being created). The following changes can be observed:

– *From a technology-led market to 'customer push'*: while the market was initially generated principally by the available technology, it is being led today by customers' needs. The result is the emergence of diversified and more complex requirements, including a strong demand for 'integrated' services;

– *Emergence of new markets*: while formerly the customers for demand TTI services were predominantly individual travellers, there is now a growing market for business applications. Travel information services are being integrated into fleet and freight management systems, and used by firms for managing their 'mobile' workforce (e.g. travelling sales representatives) as well as geo-marketing. This market offers good opportunities for TTI providers to sell personalised components (aggregated data contents, special services, applications, interfaces, etc);

– *From an 'equipment' to a 'service' market*: the first onboard equipment for delivering TTI consisted of simple (static) navigation systems. These soon revealed their limitations, due in part to their lack of user-friendliness and partly to the difficulty in updating the road network information. Demand now exists for services offering dynamic information on traffic conditions and the state of the road network. The availability of new communication channels (GPRS and UMTS) and increased coverage of the RDS/TMC service has provided the basic elements to meet this requirement. New user terminals now on the market, with better displays and greater processing power, are also extending the potential for TTI services.

– *A move towards interoperable applications*: it is now becoming more common for TTI services, such as route guidance, to be received from a source external to the vehicle and delivered to the user by a variety of different media, including on-board units, the Internet and also personal mobile devices (which therefore need to be capable of locating the user).
– *Tailoring to specific market segments.* Interoperable services of the kind described above have the advantage of being easily adapted to individual needs. They can be received via various types of terminal and integrated in different kinds of application. It seems probable that this trend will be reinforced in the near future by the expected convergence of mobile phones and portable computers. If in-car architecture becomes more open and modular, as seems likely, it will permit the integration of personal devices within the on-board platform.

3.2.3.3 *Technology developments*

Important progress has been made with the communication technologies required for advanced TTI services (GSM, DAB, microwave links, etc.) as well as the related protocols such as WAP, XML, TCP/IP, TPEG and DSRC.

Digital Audio Broadcasting (DAB) channels and broadband communications are particularly suited to TTI applications because of the mobile reception offered. The integration with GSM and positioning technologies like GPS makes it possible to provide both broadcast and interactive TTI services, and also to filter messages for given geographical areas. A wide range of DAB-based applications, from simple traffic information to dynamic navigation tasks, have been investigated in various European projects. Tests have also been carried out with different levels of service, from a 'basic' level (free of charge) to a 'premium' customised service (on a pay per use basis) in order to test their commercial acceptability.

Exploitation of the TPEG independent protocol (for DAB and Internet) for the transmission of traffic and travel information by digital broadcast systems has also been explored with a view to standardisation.

3.2.3.4 Market regulation

In parallel with the expansion of TTI services, there has been growing awareness of the need for regulation of certain aspects of this market. A significant step was the publication by the European Commission (EC) in July 2001 of a Recommendation (European Commission, 2001, (2)) which had the aim of:

(a) Facilitating the development of TTI Services in Europe;

(b) Encouraging Member States to set up regulatory frameworks for such services;

(c) Encouraging the private sector to play an active role in network monitoring.

Two main ways were envisaged of improving the availability of traffic and transport data. Firstly, by establishing a framework with a clear set of rules for granting access to existing transport data by service operators, and secondly by allowing private operators to install their own monitoring systems, again within clear rules.

Member States were given two years in which to report progress on these activities. Although this period expired without tangible progress being made, the initiative was a sound attempt to deal with a serious (and complex) problem and is certainly worth following up.

Recognition of the relevance of the issue was the Position Paper on TTI produced in March 2003 by the Euro-Regional projects (part of DG TREN's TEMPO programme). This endorsed the EC's recommendation and suggested that project partners 'should set up easy access points for service providers to access the data currently available'. It was based on the consideration that:

- The Euro-Regional projects had already decided to establish a network of traffic centres (based on the DATEX standard);
- This would facilitate cross-border traffic management;
- It was not, however, enough just to foster international TTI services, since service providers could not base their business on a series of bilateral negotiations;
- The most reasonable way to initiate the necessary co-operation would be to create easy interfaces to service providers.

3.2.4 *Issues*

While the vision for future TTI services is clear, it is rather less evident how progress in this direction can best be achieved. The principal obstacles seem to be in regard to the organisational and legal framework rather than technical factors. Experience suggests that the necessary boost to TTI services will come from opening up the market. But if a large-scale market is to be created, favourable conditions must exist for private 'value-added' TTI service providers to meet the needs of given market segments. So how can this be done?

As noted in the ATLANTIC survey, one of the major difficulties is that of harmonising the positions of the stakeholders involved. The travelling public often has very high expectations, but is rarely prepared to pay the real costs of providing such information. There is also a need for clearer relationships between service operators and providers regarding data collection, ownership and exchange if a truly open TTI market is to be created. Finally, especially if multimodal and cross-border services are to be operated, an overall framework needs to exist at European level to ensure data availability between different modes and countries. The main issues examined below therefore concern:

- Ways of improving data quality;
- Ways of opening up the TTI market;
- Ways of establishing a common high level approach in Europe.

3.2.4.1 *Monitoring the transport network*

The current coverage and quality of network monitoring is, in general, very poor. To favour such activities by operators, a number of technical and organisational barriers need to be broken down.

A useful start would be an agreement at the European level on a suitable common standard for the monitoring of all major transport networks, allowing time for all countries to reach the agreed level. Such arrangements are especially necessary in the case of deregulated systems. Where transport services run under license or franchise, operators should be asked to publish certain information as part of their service contract. This could include not only timetables and network conditions, but also real-time traffic data and information on disruptions, such as roadworks and closures. Good practice should be promoted. The work of the European ITS Framework Architecture could help in this activity.

3.2.4.2 *Clear rules for the information market*

Monitoring the networks is only a first step – the information then needs to be made available to the public. For a TTI market to be efficient, information must be exchanged between various operators (public and private). A prerequisite is that the rights of ownership to such information should be clear. There is currently no common basis for such rights in Europe. The EC Recommendation 2001/551/EC was a move in this direction, but progress has so far been limited. At the very least, transport operators in the same public sector, or acting as concessionaires or licensees, should be encouraged to make available their information directly to service providers or to telematics platforms which 'pool' TTI data. The information would not necessarily have to be provided free of charge. Part of the revenues generated by TTI services delivered by Value Added Service Providers (VASPs) could be used to remunerate the efforts of transport operators to monitor their networks. What is important is that the conditions should be clear, stable and equal for all operators. (The activities of Transport Direct in the UK could be taken as a good example).

3.2.4.3 *Efficient distribution of information*

Once the rules on rights of use have been established and the information exists, it still has to be made available in some form. This means that suitable standards (for both publication and exchange) must be defined and agreed. Information from multiple information sources will need to be combined in various forms (according to specific user needs) and delivered through multiple delivery channels.

It is not yet clear which would be the easiest (and most practical) way of speeding up the process. The normal answer – to define a common set of standards, products and data formats, and to push all parties to comply with them – is unlikely to be the most efficient. It would take too long and encounter too much opposition. The data model is probably too complex for such an approach to be viable. Moreover, VASPs need the maximum freedom to define their products in relation to the needs of their target market. Another approach would be to consider the Internet as common ground and publish the information on the web. A VASP could then find a way of obtaining the data needed, combine it as required and present it to their users. Even with this approach, some basic agreements on individual data models would be needed.

3.2.4.4 *The European panorama*

Since the TTI market necessarily has a European dimension, the commercial and legal picture across Europe must be fully understood. An analysis of the requirements for facilitating the creation of ad hoc brokers in Europe would therefore be very valuable. With the widespread use of e-commerce for reservations, ticketing and payment of services during a trip, there may be a need to review the practices adopted by the various market participants. In addition, since transport regulations are likely to affect the development of telematics services, the complex pattern of regulated and unregulated participants in different European countries and different transport modes need to be considered.

3.2.4.5 Understanding user requirements

While previously only generic services were offered, a more mature situation is developing in which specific market segments are being targeted. It is already clear that the market for 'navigation only' services is limited and is slow in creating revenues, while integrated services (such as travel plus tourism information) are more appealing. It is also evident that a large TTI market exists for business applications (the management of the 'mobile' workforce). As yet, however, a satisfactory market segmentation for TTI services does not exist.

There is therefore a major need to analyse the requirements of different kinds of traveller and to survey core segments in order to assess the type and level of detail of the information required and, equally important, to identify cost-effective ways of providing it. Such a survey would enable the European Commission to understand, for example, the potential of TTI services for supporting multimodal travel across Europe for business or tourism. Another specific segment consists of non-European nationals, who are likely to have special needs when travelling in Europe. Last but not least, it is important – especially in an ageing Europe – to understand the requirements of elderly or disabled travellers.

Much research has already been done, but a focus is now needed on what is required to develop brokerage at a mass level. However, before this is achieved, there is a further issue to be resolved. It regards the question of personal privacy which is raised by the use of personal profiles and the location of individual vehicles and devices.

3.2.4.6 Establishing minimum service levels

It is now commonly agreed that TTI services can help to provide higher service levels as well as greater safety and efficiency on the transport network. Actions are therefore needed to encourage service improvements, for example by getting transport operators to voluntarily set targets and to give priority to customer satisfaction. The practice, for instance, of giving the expected travel time for a traveller entering a given part of the network could help to raise the service level. Moreover, this would have a series of positive implications for safety and efficiency, but could only be based on extensive network monitoring. The Commission should explore, with the operators, whether practices such as these could become commonly adopted in Europe, at least in the TEN domain.

3.2.4.7 *Human-Machine Interface*

While mobile travel information has an enormous potential market (including all drivers with navigation systems plus the owners of smart mobile phones), there are Human-Machine Interface (HMI) design problems in presenting detailed, and really useful, travel information. Typical displays have very small viewing areas and the usual input modes are also limited, e.g. numeric keypads for the composition of text messages. With GSM and DAB data transmission, mobile units can produce partial candidate routes from map data, public transport timetables and live information about public transport vehicles, but better solutions still need to be found for the presentation as well as ways of preventing driver distraction.

Guidelines were developed in the 4th FP for the design of the user interface for TTI systems, regarding the layout, icons, presentation, ergonomics, sequencing of information, and so on. This was followed up in the 5th FP with the specific consideration of HMI concerning:

- Fixed locations (kiosks, displays at stops, home PCs, etc.);
- Mobile devices (on-board displays and handheld devices);
- In-vehicle and roadside displays (in-car terminals, VMS, etc.).

One result to emerge was the desirability of harmonisation of the style of presentation to have a recognisable 'look and feel' with different media and environments (on-board systems, handheld devices, etc.).

One approach for car systems is to use 'vocal' controls, but speech recognition technology still lacks robustness with respect to variations in pronunciation (most research has involved American English). Work is being carried out to make systems less sensitive to linguistic factors, and more suitable for European languages. Prototypes have been developed to provide multilingual vocal access to applications and information sources on the Internet (public and private service providers) accessible by means of kiosks, standard telephones and smart wireless devices.

3.2.4.8 Open telematics platforms

One of the reasons that TTI service providers in existence today find it hard to satisfy travellers' needs is the difficulty in obtaining high quality information. The data they require comes from many different sources (transport operators, tourist services, etc). This is typically raw data on the local traffic and travel conditions, which means it is often non-homogeneous and uses non-compatible reference networks. The result is the need for a large amount of data elaboration, which they can rarely afford. An alternative is for them to collect the relevant information directly (see the above examples on 'vertical operators'). While this is possible for some networks (e.g. the motorway network), in general it is too expensive and time-consuming, especially for urban road networks and public transport services.

One solution to this problem would be for public authorities to encourage the creation of telematics platforms to serve as 'pools' of transport data relating to a specific geographical area. This would be in the interest of service providers who would have a single point of access, to transport operators who could share their data and increase the visibility of their services, and final users, who would have easier access to multimodal and multi-network information. The centralised data management would also make it possible to offer guarantees to the quality and availability of the information supplied.

Several such platforms have been set up as part of EC projects. As part of the 5th FP, a platform known as TITOS was set up in Turin in 2000. This is still operating, and combines data on different types of transport in the city to provide real-time information on bus and tram arrivals, private traffic and parking availability. It is made available to the public through various channels (e.g. at bus stops, on-board and VMS), as well as personalised services which can be accessed via the Internet and mobile phones.

A more detailed assessment of the implications of setting up such platforms is made later in the section 'Future Opportunities'.

3.2.4.9 Lack of an open 'overarching' framework

A serious constraint to the creation of an open TTI market in Europe is the lack of clear rules governing competition. This risks slowing the growth of Europe-wide multimodal services, leaving users with limited information on alternative options, and the tendency to make the easy choice - the private car!

The survey (European Commission, 2002) published by the ATLANTIC project reveals the significant variations in regulation that exist from country to country. Although public organisations are almost everywhere supposed to make their data available free of charge, currently only nine of the 25 Member States have provision for data collection by the private sector. A feature lacking everywhere is the existence of service evaluation guidelines.

It should be pointed out that even a legal obligation to make traffic and transport data available does not necessarily resolve the problem. Very often the operators of transport networks are not in a position to (or willing to) distribute their data. The result is that in only a few cases is such information actually accessible to service providers; this situation is even more critical where real-time data is concerned.

Most existing TTI services have been developed by transport operators seeking to support and market their own operations. This approach, however, tends to limit the penetration of multimodal personal transport services. Across Europe, the extent and sophistication of TTI systems and related services (booking, payment, assistance, etc.) is still very variable. While some countries already have some highly advanced services, others are at a very basic or incomplete stage. This is a further constraint to the development of services for multimodal cross-border trips.

3.2.4.10 *Interoperable multimodal information systems*

Prototypes of many of the tools needed for interoperable multimodal TTI were already developed in the 4th Framework Programme (FP), and some of these were tested in small or medium scale demonstrations. They included services for language-independent traffic information exchange systems and advanced driver warning systems. A European reference data model (Transmodel) was produced for public transport operations.

These results were built upon by 5th FP projects. Work was carried out on establishing common mechanisms for exchanging data between content owners for different modes (bus/tram/metro, rail and road), and also on identifying the organisational and strategic issues hampering multimodal travel. Innovative paradigms for the design of open, distributed and networked tools for TTI were developed to favour a new class of just-in-time, interactive, value-added, map-based and personalised travel services. A prototype integrated toolkit was created to access information sources using XML/Java and view them on wireless hand-held devices.

In the 6th FP, as part of the eSafety initiative, the project IM@GINE IT is combining previous results and agent-based technology to create a universal platform covering urban, interurban, and cross-border areas. This platform will serve as a single access point through which the end user can obtain a set of TTI services everywhere in Europe.

3.2.4.11 *Data exchange requirements*

The data exchange systems developed at a European level have until now mainly been based on DATEX specifications, which impose the use of the ALERT-C location referencing system. This choice for the exchange of road information in the interurban environment was made when no other solutions were available. Today, the context has changed.

Data is being exchanged in both the urban and interurban environment. Traffic information systems are evolving from road-only information to multimodal information. Location referencing systems, which previously used just a few mono-mode pre-coded locations, are moving towards multimodal 'on the fly' codification (e.g. AGORA or ILOC, a map-based location reference method based on X-Y co-ordinates associated with descriptors and location types). As a consequence, the methods being used in current data exchange systems show that there is still substantial uncertainty about the most suitable method for intermodal data.

3.2.4.12 *Privacy and confidentiality*

The growth of highly efficient personalised TTI services carries the risk of negative side-effects regarding privacy. When planning a trip, travellers make known to the service provider their destination and often other information. If location-based services are requested, or mobile navigation employed, the traveller must by definition be located and often tracked. To provide an efficient and rapid service for frequent users, a service provider may also wish to build up and store personal and trip profiles. Travellers can then be 'alerted' only when emergencies are likely to affect them, and individually tailored routeing suggestions given.

In all of these cases, a service provider can build up a detailed picture of a customer's travel habits. As a result, there exists the danger that this information could be used improperly, posing a threat to confidentiality. This makes it important to extend existing regulations on personal data protection to cover TTI services and to find ways of creating personal data files without harming privacy rights (and without obliging the user to complete complex procedures each time a service is used).

The scope of national legislation, in this respect, should be explored. In parallel, research on technology and applications is required in order to find practical ways of protecting personal data. One possible solution could be the combination of smart card technology and wide band communication, or the extended use of 'nomadic systems', which would allow personal profiles to be stored only in equipment which is under the full control of the user.

3.2.4.13 *Safety/security aspects*

TTI services can have a positive impact on safety (as they may encourage travellers to choose safer modes/routes), but they can also have negative impacts resulting from distraction while using a terminal (especially while driving). For a more detailed discussion of this issue the reader is referred to the section on HMI. In the specific context of TTI, recommendations will need to be formulated to reduce the risk of distraction leading to accidents. This will be especially critical (though hard to regulate) if the use of nomadic systems becomes widespread.

The issue of personal security is parallel to the privacy issue. Since users of TTI services can be traced, this can be used both to their benefit (in the case of emergencies) and detriment (if data is used improperly).

3.2.5 Future opportunities

Three important aspects of future TTI services are described here. These regard:

(a) The opportunities offered by new technologies;

(b) The potential of open platforms as 'collectors' and processors of transport-related information;

(c) The social benefits of TTI used as a transport management tool.

3.2.5.1 *The potential offered by new technologies*

If in-car architecture in the future becomes more open and modular, as seems very likely, it will permit the integration of personal devices within the on-board network (this is already true for mobile phones, and will soon apply to more complex systems). Such a development would enable the use of certain in-car components (like the location subsystem, loudspeakers and microphones, dashboard display and some commands) to supplement the capabilities of the so-called 'nomadic systems' (e.g. PCs with high processing power and a portfolio of user applications). This will be made possible by the availability of efficient, low cost wireless connections within the car (today implemented via Bluetooth technology). The potential benefits for the TTI market are evident.

A further opportunity regards the possibility offered by the use of images (video streams or pictures) in TTI services. The use of GPRS and UMTS rather than GMS gives greater communication bandwidth which could be used to make applications more attractive, e.g. visual information (MMS) rather than text (SMS). This would for example permit visual navigation with pictures of important milestones to supplement maps (e.g. turn right after you see this monument) or substitute descriptions of 'Points of Interest' (POI). Real-time images of traffic queues could back up the textual information. Such applications could well become inherent parts of a navigation application. While it is impossible to predict if these would capture a large-scale market, they would certainly meet the interests of big market players (communication carriers) and could dramatically change the 'content' of TTI services.

Another significant development with potential for considerable impact on future TTI services is the GALILEO initiative. In addition to the greater precision this system will provide with respect to positioning, an important benefit for commercial services will be the service guarantee. (The issue of radio navigation is discussed in more detail in section 7.2.)

3.2.5.2 *The advantages of creating 'open' TTI platforms*

It is assumed that a number of different business models will operate in the future, and that these will be based on the concept of an 'information market'. In other words, traffic and transport information will be exchanged between operators under market conditions, although this will occur within some form of regulatory framework.

While basic transport information is likely to be provided free of charge by transport and infrastructure operators, a growing number of added-value services will be offered on a commercial basis by private service providers. Travellers will choose the appropriate level of service according to their needs and willingness to pay. 'Niche' services will cater, for example, for the particular needs of business travel, city breaks, special interest tours, etc. and will be especially useful for travel in foreign countries and trips involving the use of several different transport modes. They will need to offer sufficient benefits in terms of time/ money saving or personalised assistance to be worth the cost to the user (Figure 4).

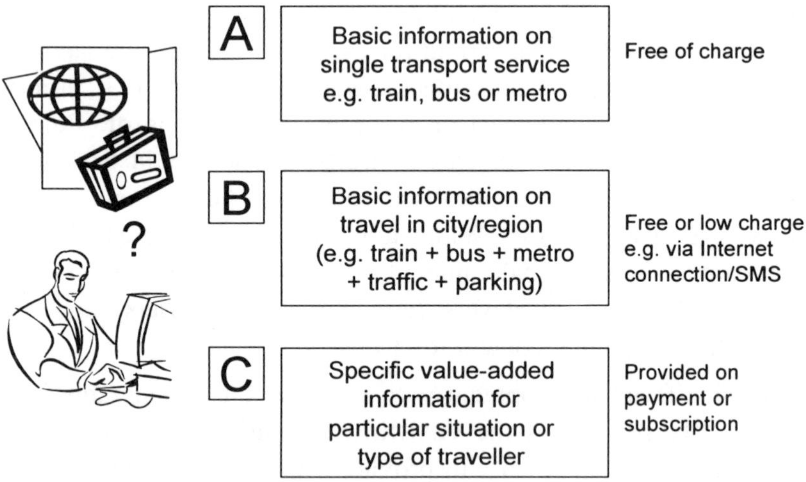

Figure 4: Three fundamental levels of TTI service

In order to personalise their services, TTI providers will collect information on frequent customers, e.g. via the traveller's use of smart cards. The creation and maintenance of profile records, will however need to be transparent to the customer, and also take into account all required guarantees of privacy and confidentiality.

The complexity of the task of collating the transport information required for specialised services means that a new type of operator is required: the Value Added Service Provider (VASP) or Transport Information Broker. This is in reality a form of 'travel agent' who will need to have access to a wide range of travel-related information (including real-time data) and also be able to process information from many different sources. Such information is likely to be non homogeneous as it will consist of raw data on traffic and travel conditions, based on different reference networks. To facilitate this task, it will be important for standards and best practice to be established at European level.

A possible structure for the provision of TTI services is represented in Figure 5. While transport and infrastructure operators cater for the interests of their own customers, the information platforms provide information for travellers in a given city or region, and VASPs offer niche services for specific market segments. The primary data flow is from operators to platforms and thence to the VASPs, who also support the platforms with data, enabling operators to retrieve further important information.

As shown in Figure 5, the VASPs in effect provide a 'bridge' between transport operators and the various market segments, and the platforms provide the first level of data processing and fusion.

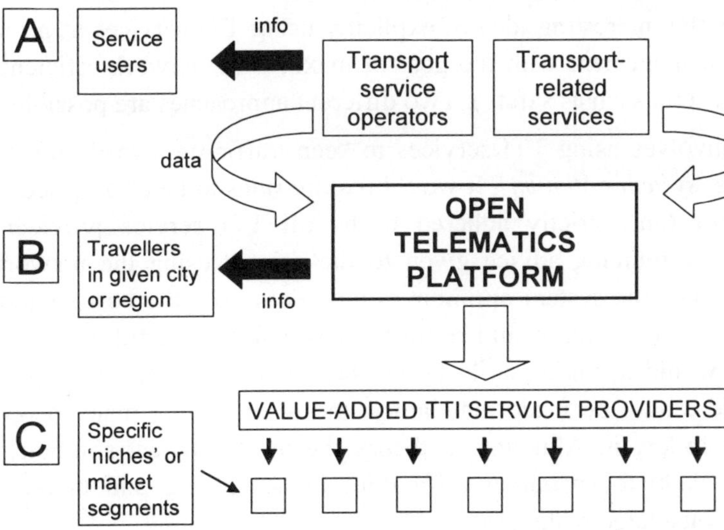

Figure 5: Data flows to ensure high quality TTI services

3.2.5.3 TTI services used as a traffic management tool

If the use of TTI services becomes widespread in the future, and these reach a large part of the travelling public, they will have the potential for generating significant social effects. Firstly, they can clearly improve the personal 'comfort' of travel (mainly through time-saving and gains in convenience). But they can also be used as a tool to improve the overall efficiency of the transport system through better distribution of mobility demand between modes and within a transport network. Successfully applied, such a policy could help to reduce congestion.

We examine this second possibility in greater detail. Results obtained in large-scale demonstrations are promising. It has already been shown that multimodal travel information can increase the use of public transport by 'promoting' alternatives to the private car (the 'hidden options'). It has also been demonstrated (e.g. in the QUARTET Plus project) that dynamic routeing can improve the efficiency of road networks, producing significant savings in average travel time.

This raises the interesting idea of explicitly using TTI services as a tool for traffic management with the goal of maximising network efficiency (or other criteria such as safety). Two different approaches are possible.

The first involves using TTI services to keep traffic on a road network close to its *'system optimum'*. It would require rules to be established by the operator (and strictly adhered to by all TTI service providers) regarding the routeing advice given to the drivers using the network. However, since the system optimum is not necessarily the best solution for each individual, this would result in drivers not being fully satisfied (the route would not necessarily be the fastest or most convenient). As TTI advice cannot be made mandatory, the result is that many people would not follow it. Although in theory the most efficient solution, it would be likely to encounter difficulties of acceptance and therefore cannot be considered really practicable.

The second approach involves using TTI services to achieve the *'user equilibrium'*. In an open market, it can be assumed that the VASPs will put the interests of their customers first, delivering the 'best' suggestions for each. Being specialised in given market segments, they will become experienced at optimising user convenience. The result will be that the network will operate at – or close to – the user equilibrium. At this point overall efficiency is lower than the 'system optimum', but all users will find their optimal route. One of the possible drawbacks is that the VASPs could be tempted to recommend unsuitable routes (roads passing through residential districts, narrow streets, etc.).

While such behaviour today is limited to a small percentage of drivers (the so-called 'rat runners'), widespread use of TTI services could lead to serious problems. However, since the equilibrium point is normally stable (repeated every day with the same demand) and therefore predictable, it is likely that a long-term solution could be found. For example, across a city road network, traffic regulations, speed limits, and traffic calming methods could be specifically designed to keep traffic to an appropriate level. In this context, the TTI services would help to keep the network functioning as close a possible to the predicted state. As a short-term measure, agreements could be sought with service providers (e.g. to limit routeing suggestions to main roads).

3.2.5.4 *Recommendations*

Over the last twenty years, research activities in the field of TTI have been numerous and have produced some very useful results. As noted in the review above, many new tools have been produced, prototypes demonstrated and studies made of important aspects such as user needs.

Although some technology issues still need to be resolved, it is evident that the main stumbling block is the lack of an open 'information market' in Europe and of standards which permit service interoperability. If the full potential of TTI is to be achieved, clear rules will need to be established to favour a Europe-wide approach.

This means that it is important that the European Commission should continue to play a proactive role in harmonising national regulations and practices regarding TTI in order to promote the following:

1. *Encouragement of network monitoring*
   - Establish the minimum information which operators should be required to publish as part of their service contracts;
   - Agreement at European level on a common standard for the monitoring of all major transport networks, with a given period of time for all countries to comply with this level;
   - Promotion of good practice.
2. *Establishment of clear rules for the information market*
   - Drawing of up clear, stable and equitable regulations regarding the rights to ownership and use of TTI data (on the lines of the EC Recommendation 2001/551/EC).
3. *Creation of conditions for efficient distribution of information*
   - Establish an agreed way of making TTI data available (e.g. via the Internet) and define any necessary common data standards and formats for allowing the publication of data.
4. *Setting up of telematics platforms (as 'pools' of TTI data)*
   - The creation of 'telematics platforms' to serve as single access points for users of transport data relating to a given geographical area (allowing operators to share data and final users to have easier access to multimodal and multi-network information).
5. *Understanding the European panorama*
   - An analysis of the basic commercial and legal requirements for the setting up of private TTI service providers in Europe;
   - Survey of regulations regarding TTI market in different countries in order to understand the picture across the whole of Europe;
   - Review of practices for use of the Internet for reservations, ticketing and payment of services during a trip.

6. *Market segmentation and analysis*
   - Definition of a market segmentation for TTI services on the basis of user needs and analysis of requirements of mass brokerage;
   - Examination of how services for vulnerable groups (the elderly, handicapped, etc.) can be provided in a cost-effective way.
7. *Agreed service levels*
   - Promotion of voluntary targets and service levels by transport service operators on the basis of a set of given criteria, e.g. efficiency, promoting customer (traveller) satisfaction.
8. *TTI services as a traffic management tool*
   - Research into ways in which TTI services can be used to promote more efficient use of the transport network;
   - Field trials to measure and evaluate the impact of different approaches.

3.2.5.5 *Future research actions required*

- Research into interoperability of information, booking and payment systems (permitting the integrated use of smart cards, mobile phones, Internet-based services, as well as on-board car platforms to allow access to personalised transport services at home and during trips, e.g. through personal nomadic systems). This calls for strong and well-accepted standards. Research should therefore consider both the technological and standardisation aspects.
- Development of 'decision support systems' for travellers. A large amount of information is needed to help them build their trip chains, but is dispersed among different sources. Support systems for online decisions will involve, for example, algorithms for multimodal trip planning operating on non-structured information, and methods for predicting transport/traffic behaviour and for interacting with the user.

Further work on search engines, especially for public transport trips where the user needs a seamless door-to-door – and possibly multimodal – solution, and a search procedure which is accomplished quickly and easily. The data itself is generally non-structured and difficult to use, so powerful algorithms are required for comparing solutions and matching with user needs. (These search engines have the difficult task of being cleverer than the user!)

Monitoring of the legal and regulatory context across Europe in order to provide a picture of the level of regulation in different countries for ITS in general and for TTI-related activities (an updated and extended form of the survey already carried out by ATLANTIC project).

Definition of user needs, with in-depth analysis of the requirements of specific market segments, with a view to developing the market for value added or 'niche' service providers.

Research into potential changes in transport demand resulting from new developments (including e-commerce and the freight domain, as well as demand management schemes, such as road user charging). This is crucial for understanding future scenarios.

Definition of a methodology for the measurement and evaluation of service quality for different types of TTI.

Development of applications which exploit the opportunities offered by innovative technologies (such as UMTS), which are likely to influence communication standards and organisational structures.

Chapter 4

Vehicles and Infrastructure

This chapter considers the development of autonomous in-vehicle systems, co-operative vehicle highway systems, the future role of the driver in interacting with intelligent transport systems and services, emergency response systems and lastly the role of ITS in enforcement.

4.1 Advanced Driver Assistance Systems

4.1.1 Vision

The vision of vehicles deploying technologies to support drivers in order to increase safety, reduce congestion and improve the driving experience was first developed in the EC PROMETHEUS programme in the 1980s (Augello, 1991). This was a major effort by the European automotive industry to understand the ways in which sensor and communication technologies could be developed and applied to best assist drivers to deal with the processes often involved in negotiating the road network, following other vehicles, undertaking complex manoeuvres and avoiding collisions in a more comfortable and relaxed environment. This vision was part of what has become known as ITS, and has been supported and enhanced through a range of EC framework programme initiatives. This is leading to the wide-scale deployment of vehicle-based market-driven ITS functions. Such technologies were initially introduced as high cost features in the most expensive vehicles, with increased market penetration following as experience was gained and costs reduced. More recently, new ITS technologies are being designed for immediate mass market introduction to recoup development costs more rapidly. This approach offers greater opportunities for early and more substantial impacts on traffic and safety.

4.1.2 *State-of-the-art*

4.1.2.1 *Driver assistance systems - their design environment*

Vehicle-based ITS systems are intended to take advantage of market opportunities and design and development is focused on buyers' needs and desires rather than to address more general traffic problems. Thus, whilst products may aim to meet both personal and public objectives, this is not necessarily the case. The approach may be considered to have a strong element of technology push towards the users, i.e. vehicle purchasers, with relatively little practical consideration given to wider transport policy objectives.

Today there is little interaction between demand from the public, those responsible for road infrastructure and its management, the road haulage industry and market-driven innovation. New innovations by industry lead to new market expectations, and determining what the various future users will really want is not always easy in such a volatile situation, as current opinions are based on imperfect knowledge of the technical 'solutions' offered. Thus, whilst it is possible to find out if a technical solution will be appreciated, the extent to which it fulfils long-term user needs remains unknown. Although, this technology push affects much ITS development, it is very prevalent in the 'driver assistance' area, where increasing automation is expected to directly support driving tasks and where only a few examples are available.

A range of technical solutions are available for many ADAS (Advanced Driver Assistance Systems) functions. The balance between public or infrastructure investment and private or users' investment will play an important if not decisive role in this choice in the future. In the driver assistance area the (sub)systems needed for one ADAS function may well overlap the subsystems needed for another ADAS function. Whilst this opens the opportunity for different functions to share costs, so far the quest for ideal technical solutions has led to system designs in which specific technologies have been selected for specific functions or user needs, with little consideration of wider applications. The reason is that additional delay involved in developing co-ordinated approaches to functions acts against the market drivers to deploy systems as rapidly as possible in a very competitive market place.

Many driver assistance system developments were initially considered as wholly autonomous in-vehicle systems. A key reason for this approach by industry was the wish to be independent of road or communication authorities, who were seen as being likely to limit the flexibility to pursue market opportunities. However, some driver assistance application developments need to use infrastructure such as beacons. In general, the role of the infrastructure is championed by industry rather than by public authorities, and successful examples of partly or fully automated infrastructure supported systems can be found on private sites such as container yards. However, critically, developments of driver assistance systems have shown that fully autonomous systems often have too little to offer the market, and even relatively modest systems rely on some form of infrastructure availability. Typical examples are ADAS solutions that use GPS (Global Positioning System) satellite navigation or lane markings. GPS is available, but practical tests show that it does not always meet the needs of some driver assistance systems and, for example, lane departure or lane keeping systems can have difficulty in handling temporary or non-standard lane markings.

More direct interactions between infrastructure and vehicles require dedicated in-vehicle systems and special, adapted and integrated infrastructure, requiring a joint effort by industry and public partners. In the case of driver assistance and automation for vehicles, public and private investments may turn out to be very large and all user needs and possible solutions must be reviewed to justify public funding. This may take some time as solutions will have to be applicable across Europe, and regional differences in geographic, weather and other conditions may lead to rather different user needs.

Only a decade ago ADAS technologies were still in their infancy, but now operational systems are available with more possible in the near future. The ROSETTA activities have been a first step towards a better co-ordination of user needs and innovation requiring:

- A co-ordination of user requirements;
- A matching of requirements and possible technologies;
- A co-ordination of deployment – a scheme for investment by industry and public partners.

4.1.2.2 *ITS application areas of driver assistance and automation*

A series of widely different ADAS functions are described below, together with comments on the results of research into technical approaches and viability.

4.1.2.3 *Automated vehicles*

Traditional public transport services best serve significant levels of demand such as may occur on urban/suburban corridors. Therefore, to enhance the relative attractiveness of public transport over the car, particularly for the journey to/from work and for business travel, a new approach is needed to address 'last' mile issues. An approach to tackling the problem of low and dispersed levels of demand is to use automated vehicles that can be shared and reused.

Groundbreaking work in this area has been done in Europe, starting with the Praxitel project in France and with more recent applications in the Netherlands. Such applications are generally regarded as interesting but still futuristic. An aspect of the French work is that they did not limit themselves to the design of a simple vehicle that might automatically be relocated (Parent et al., 2002). They also looked at features such as automatic parking that could make the concept more attractive to users. Even though deployment is currently limited to low speeds and segregated tracks, applications and trials have provided insight into the functional requirements. Technological development will, with time, make solutions more affordable and may enable vehicle speeds to be increased above the present practical level of about 15 km/h. Fully autonomous vehicles have application limits which may be overcome by greater integration with infrastructure as discussed in section 4.2 (Vöge and McDonald, 2004)

4.1.2.4 *Docking buses*

Automated docking systems, which let buses stop just a few centimetres away from the platform, expedite loading and unloading as well as offering a better service, particularly to those with special needs. This application is a spin-off of lane keeping technologies. In the past, mechanical solutions have turned out to be rather unsatisfactory, but current electronic versions using a video base, or the magnetic sensor supported systems, have been shown to work well. Additional design features on systems such as the Phileas bus (Siuru, 2004), where all wheels of the bus can be steered, enable the bus to move sideways at sharp angles.

4.1.2.5 *Emergency vehicles*

In the American IVI programme (Intelligent Vehicle Initiative) the need for driving support for emergency vehicles that have to operate under very adverse weather conditions like snow and fog has led to the development of a lane keeping system for snow ploughs, ambulances and police vehicles. Two location technologies are used; where GPS signals are received, the information is combined with triangulation of dGPS information to achieve a position accuracy of 3 cm. Where GPS reception is poor due to specific geographic circumstances, such as city canyons or roads with an umbrella of wet leaf trees, magnetic tape in the roads can be used for guidance. The system uses accurate maps to show the driver the delineation on a head up display. The map information is also used to determine the speed, especially relevant when approaching a sharp curve. The system is now operational on a range of emergency vehicles in Minnesota (Kwon et al., 2003).

The very special situations for which this sort of driver support has been developed make it a rather unique solution; it is expensive and needs a trained driver. So far even in Nordic European countries, road operators have found these applications too impractical and/or too expensive.

4.1.2.6 *Garage management*

Inside the garages of fleet operators, trucks or buses are moved during the daily maintenance routine. This handling of vehicles is costly in labour time and technologies exist which could enable the automated management of vehicle movement in such a specialist and controlled environment.

4.1.2.7 *Intelligent Speed Adaptation*

Intelligent Speed Adaptation (ISA) (sometimes known as Intelligent Speed Advice) was originally designed to help drivers not to exceed the prevailing speed limits in urban areas where the effect of a collision with a pedestrian or cyclist may be minor at speeds below 30 km/h and fatal at speeds over 50 km/h. The system can be expanded to work on all roads. There have been extensive tests in Europe which have identified that a system could be developed that is both effective and attractive to users. High levels of enforcement and penalties for speeding would be key market drivers, or governments could pay for and legislate for system introduction (Stefan, 2003; Biding, 2004).

Autonomous systems can never be up to date with their information on prevailing speed limits, and would need infrastructure support for vehicle location and speed limit communication.

Therefore, it is interesting to note that in Japan the VIC-network of beacons along the road will be used to communicate actual, prevailing speed limits to vehicles. Vehicles that have a VIC-link in their navigation system not only show that speed to the driver but also warn against speeding.

4.1.2.8 *Mayday e-112*

A recent development is that of E911-Mayday and e-112-CGALIES emergency call handling to automatically inform emergency services of an accident for more rapid and effective response (CGALIES, 2002; Paola et al., 2003). Take-up by car owners is very impressive and millions of vehicles are already equipped. Although the services themselves have no direct impact on the issue of driver assistance or automated vehicles, the related needs of accurate digital maps has a much wider value. Several solutions use information from the cellular network operator; this is not sufficiently accurate in many circumstances and such solutions are regarded as temporary. The use of both GPS and map matching can increase accuracy substantially.

4.1.2.9 *Obstacle warning*

Radar, video and laser systems have all been studied, either singly or in combination, to identify obstacles and warn the driver of possible collision (Nico and Klaus, 2004). The systems must identify the obstacle in the context of the normal physical characteristics associated with the road ahead and so are very complex.

To date, such systems have been used on special types of vehicles like snow ploughs, ambulances and police vehicles. Radar is used to put the lane and road edge boundaries on a Head-Up Display for the driver. Obstacles on the road ahead also appear on the Head-Up Display and the driver is warned. The resulting control and accurate network mapping for precise lane support has other potential applications for road safety.

4.1.2.10 *Adaptive Cruise Control*

The automotive industry developed Adaptive Cruise Control (ACC) which is now in volume production (Marsden et al., 2001). These systems use radar to measure the distance to the vehicle ahead. The distance between the vehicles is controlled to keep a constant time gap selectable by the driver. The control system adjusts both the throttle and brake systems to maintain the correct separation between vehicles. ACC has been well received by the public who see it as a comfort system. This autonomous arrangement is a good stepping-stone towards ISA and curve ahead warning services.

4.1.2.11 Navigation systems

Navigation systems have become commonplace in new cars. These systems use satellite navigation plus map matching. Their success in finding routes can mislead users into thinking that the systems always know exactly the location of the vehicle. That is not the case. Vehicles are navigated with the help of maps that may be tens or even hundreds of metres wrong as well as with the help of GPS information that also can be inaccurate or sometimes unavailable. These inaccuracy problems are overcome by clever software in the navigation program. That program combines the last calculated position with new GPS information and then checks on the map where the vehicle might be. So if the GPS information would place the vehicle away from the road the map matching mechanism corrects this mistake. However, this mechanism may fail when there are rather similar alternatives on the map. If the navigation system finds that map matching is not possible it will backtrack to the point where it could choose between options and try to match the lately collected series of GPS locations with the alternative route and thus find the better match. This option of self-correcting the position by backtracking works for navigation but would not be adequate for other systems such as ISA. However, it is a valuable step in helping users accept in-vehicle systems (Hu et al., 2004).

4.1.2.12 Ports

The need for accurate location information in harbours is recognised and various technologies are being tested to develop operational systems for port management. In this highly technical environment, investments for automation are acceptable when they improve safety and efficiency, and the market is developing rapidly.

Such technologies may be directly or indirectly relevant for road traffic once matured and large-scale production becomes possible. The possible overlap of technology solutions between waterborne and road transport is hardly recognised by the authorities.

4.1.2.13 Guided Buses

Automated lane control on a public bus can enable a narrow lane to be used, thereby reducing costs of new construction or sometimes enabling an additional priority lane to be gained in existing infrastructure. Such buses are of a dual mode type; the driver can also run them under regular manual control whenever the buses use 'regular' public roads. Both video-based technology using special painted reference lines in the middle of the (bus)lane, as well as magnets in the road and magnetic tape systems, can be used for guidance. As these systems offer haptic support, no 'steer by wire' is needed.

Older mechanical systems have been introduced in Japan, Germany and elsewhere, but lack of comfort and high maintenance costs were problems. Electronic versions operate more smoothly, as in the example of the route between Minneapolis and Saint Paul where buses bypass congestion on the freeway by driving on a hard shoulder which is only a few centimetres wider than the bus (Alexander et al., 2005).

Road operators in Europe generally welcome this application, but the fragmentation of the public transport network is seen as a problem.

4.1.2.14 Run-off-the-road accidents

More than ten percent of the fatal accidents on rural roads occur as a result of accidents in which the vehicle has run off the road. Lane keeping support has been developed to reduce the likelihood of such accidents. Operational versions are already on the market and this is discussed further in the section on Lane Departure Warning.

4.1.2.15 Truck efficiency

To reduce the costs of trucking, platooning of vehicles following at close distances with an electronic tow-bar has been developed, with savings in fuel being greater than the costs (Bonnet and Fritz, 2000). This platooning also allows only one driver to control the whole platoon, thus also saving driver costs or workload. The mechanism may look simple, but sophisticated programs are needed to ascertain that all trucks in the platoon follow the same course when driving through curves.

This financially driven technology push may provide trucks with control facilities that can make the introduction of other driver assistance systems simpler and cheaper. Whilst costs across a platoon are reduced, those for the first vehicle are increased. Thus, the structure of the trucking industry, particularly the number of different truck operators, becomes an issue for widespread uptake.

4.1.2.16 Tunnels

On entering a tunnel, drivers tend to move towards the middle of the road and away from the walls. This can lead to dangerous situations and accidents in tunnels can be more severe because of the confined area with less visibility. Much infrastructure is already deployed to improve tunnel safety and lane keeping support could help as an additional preventive measure. Many road and tunnel operators recognise the potential of the solution, but the scale of the application requires it to be a spin-off solution based on a larger scale application (Takuya, 2003).

4.1.2.17 Warehouses, container yards and amusement parks

Fully automated transport is used in warehouses, container yards and amusement parks. These bounded areas are under the control of only one party and savings on labour costs have led to reliable products that function well. These systems can be learning grounds for technology development. However the situations differ so much from those on the public road the success of these systems cannot be a guarantee of their general application.

4.1.2.18 Lane Departure Warning and Lane Keeping

Current operational systems in this area range from autonomous video based lane departure warning systems that emit a noise when a vehicle tends to cross the white line marking the edge of the lane, to infrastructure supported systems that allow driving in whiteout conditions thanks to the projecting of the road's geometry on a head-up display. The relatively simple video lane departure warning systems are expected to serve many needs. They could be used to reduce the number of run-off-the-road accidents, avoid tunnel collisions, and reduce stress at narrow roadwork stretches (Alkim, 2003).

As a means of warning drivers against leaving the road, the systems function well provided the delineation markings are visible. This latter requirement is normally not a problem on long distance highways where the warning is most needed. Tests show a distinct difference in use by truckers who tested LDWA (Lane Departure Warning Assistance) systems: they were used 75 % of the time on long distance freeways and 60 % on lesser rural roads. The warning system was found to be less appreciated on narrow lanes; professional drivers prefer to manoeuvre their vehicle without a LDWA support as it warns of lane departure too often in these circumstances. This implies that lane departure warning systems should use haptic steering or automatic steering rather than just warning. However, on sections with, for example, temporary yellow lane markings overruling the still existent white markings, simple autonomous video systems were found not to be able to cope. Also, drivers who tested LDWA on cars in a version which aims at rather perfect lane keeping did not appreciate it very much as a product, as it made driving 'dull'.

Both the more sophisticated automated snowplough systems and the public transport bus systems in Minnesota are based on the same technique (Alexander et al., 2005). GPS navigation is supported by triangulation of the differential or correction signal available from three or more base stations. Magnetic coded tape in the road surface is used where reception is a problem. As snow ploughs may have to operate under whiteout conditions the geometry is displayed on a head up screen. This use requires additional driver training. For bus lane keeping, haptic steering is the preferred mode of operation when buses use a very narrow hard shoulder to bypass congestion on the freeway. The lane is narrow but the driver has full visibility. Simple warning and full steering control are the other options. Other systems that test for drowsiness may also be used to address the run-off problem. These systems can be based on erratic steering behaviour or on the monitoring of eye movements. First systems are on the market which monitor the behaviour of the driver and alert the driver when he or she does not pay sufficient attention to the driving task. However, they will not warn against running off the road while making phone calls or brewing coffee!

From today's operational system, it may be concluded that there is no simple solution that works for all the lane keeping problems. On the other hand, good solutions are technically possible, but their costs may be prohibitive. Most important is the need for infrastructure support which will turn out to be the decisive factor for the future.

4.1.2.19 A tale of research

The work on LDWA shows an interesting historical development of research targets. When capacity became more and more of a problem for freeways in the Netherlands, the idea evolved of changing 3-lane carriageways into carriageways with 4 narrower lanes. The maximum speed would at the same time be reduced from the regular 100 or 120 km/h.

The concept was based on line marking used in indoor sports halls, where it is common to work with different colour markings for different types of sport – a system which does not cause confusion to users. However, the use of different colours for lane markings to indicate the 3- and the 4-lane situation on the same carriageway was not considered acceptable to the highway authorities. Therefore, dynamic lane marking using LEDs (Light Emitting Diodes) to show a white line was developed. After a series of trials, in which more than one prototype failed, operational systems became available. The first operational tests of these LED-line markings used the hard shoulder as a traffic lane in peak hours. In this case, to separate the hard shoulder from the 'regular' open lanes the LED showed an uninterrupted line. When the hard shoulder could be used as a traffic lane the LED showed a broken line instead. This arrangement was found to work as an operational test of the use of dynamic marking, although the idea of changing 3 lanes into 4 more narrow lanes was not favoured by the public.

The result of the LDWA tests with truckers showed that following the lanes was no problem provided that it was done at the officially posted rather low maximum speed. In practice, traffic tends to drive at much higher speeds, and that makes lane keeping in a narrow lane more difficult and a driving burden. The two approaches to a solution would be to either control speeds so that drivers could cope with reduced lane width, or to adopt a technical approach to control vehicles at the higher speeds adopted by drivers. The first is the simplest answer, but not one adopted in an environment of technology push. However, a change from 3 regular to 4 narrow lanes could still be achieved with the help of ITS, not in the form of LDWA, but by modern speed enforcement to restrict traffic to the lower posted speed. ITS solutions are already available for such enforcement. Clearly, this is a rather simplistic approach, but illustrates the need for solutions to be driven by clear user needs rather than by technology opportunities.

4.1.2.20 VII – What's in a name?

In the USA, the IVI (Intelligent Vehicle Initiative) programme (Hanowski et al., 2002) evolved into VII (Vehicle Infrastructure Integration) (U.S. Department of Transportation, 2005) with the recognition that autonomous vehicle based systems alone would not lead to the user/system benefits available from the integration of information. The VII plans received a boost by the acceptance of a single vehicle-beacon concept after many years of debate about frequencies etc.

Today's VII plans centre on an arrangement where vehicles communicate with each other as well as with beacons. The idea is that information is exchanged between vehicles and beacons when in short range communication reach of each other and that vehicles further pass on information to each other so that those which are out of direct reach of a beacon share the beacon's information. This sort of arrangement opens up a wide range of ITS opportunities. There are several initiatives in this area. For example, the German automotive industry successfully trialled a hopping arrangement, a means of communicating well known in radio-communications. These initiatives have a historic forerunner in the form of the Pro-Net work of the PROMETHEUS project, rightly seen as one of the major projects in ITS history. In Pro-Net the possibilities of using communications between vehicles were studied extensively, especially the need for timing arrangements, for positioning coding and for control co-operation.

4.1.3 Issues

The concepts of advanced driver assistance and autonomous vehicle systems, together with many of the technologies, have been available for many years. Applications of all the functions have been demonstrated but implementation has been slow.

Location is at the heart of all the functions, whether it is absolute location, the relative location to other vehicles or obstacles, or location with respect to the road network. The difficulties in bringing systems to the market stem from the problems in ensuring accuracy of location and reliability of that accuracy relative to the application. Thus, navigation systems are now widespread because of the quality of GPS and map matching systems and because the penalty of occasional system failure is small. Where system failure introduces additional risk, the systems have not been introduced. Therefore, key issues for further development of systems, and the significant overall benefits which may accrue are:

(a) The extent to which fully autonomous systems can be made fault-proof, to a level which is acceptable to users who would purchase such systems and to governments who would legislate to support their introduction and operation. Industry may be able to overcome some of the problems through the development of non public vehicle-to-vehicle and vehicle-to-roadside systems. Such network systems could enhance various autonomous ITS functions and would be market driven without the need for substantial governmental funding.

(b) The extent to which infrastructure support can be used to enhance autonomous system functions to a level which becomes acceptably fault-proof. This introduces additional issues of where liability lies, how joint commercial, financial and governmental economic cases can be combined, and how the timescales and decision processes of government and industry can be reconciled.

(c) Ways to bring together industry and governments to enable public and private funds, legislation, and enforcement to be better focused on achieving a joint approach which will encourage drivers to purchase and use systems which benefit them as individuals and society as a whole. This will involve a more coherent view of the ITS functions and technology needs.

Location accuracy, relative and absolute, and the use of appropriate technologies remain key issues. ITS functions at or near implementation are shown in Table 4 with an 'L' for those which require lane recognition, or with an 'X' for those that do not.

Note that the functions all rely on location and each has its individual location accuracy and digital map requirements. Normal development will produce a range of these digital maps with varying detail and accuracy. Individual system development can make concessions on location accuracy to keep costs low and enable progress to be made in deploying them. By determining the requirements of each function for location and accuracy the correct needs can be determined. The need is to develop an agreed plan showing the priority of each function, the accuracy required of the map, location system and the choice of technology. By taking positive action at an early stage in the programme to develop functions which have Government priority, it should be possible to take advantage of the most accurate location technologies which would then service the less demanding applications, and thereby saving costs.

Table 4: ITS Functions at or near Implementation

Functions / Technology	Mayday	ISA	Collision Warning	Road Curvature Warning	Lane Departure Warning	Snow Plough	Bus Lane Support	Ambulance, Police, Fire, De-icing vehicles in fog	Roadworks	Road Bottlenecks	Bus Docking
Triangulated dGPS			X		L	L	L	L	L	L	L
Magnetic Tape			X		L	L	L	L	L	L	L
dGPS		X									
GPS (GALILEO)				X							
GPS with 3 tracks	X										
Digital map with lanes			X		L	L	L	L	L	L	
ADAS Digital Map	X	X		X							
Haptic Steering							X		X	X	X
Radar			X			X		X			
Haptic Throttle		X		X							
Head-Up Display						X		X			
Electronic Rumble Strip					X						

Lane Support is required for narrow bus lanes. However, for successful application this function requires the most accuracy for both location and maps. To achieve the required accuracy Triangulated dGPS (differential GPS), Magnetic tape and very accurate detailed digital maps are required.

4.1.4 *Future opportunities*

From the listing of driver assistance systems it is clear that they share the need for reliable location information as well as the need for up to date support. (New roads will be opened, and road works can affect the geometry on a temporary basis.) A range of technologies have been trialled for lane keeping, the most demanding function, and working solutions are available. Whilst these are expensive, it is anticipated that the costs will fall rapidly with widespread use and by sharing use with other different ITS applications. However, so far not even the use of satellite navigation is certain, let alone the ways in which supportive measures can assure additional location information. Differential GPS may work well in many cases. Additional satellites may be used to improve the GPS service as is now the case in metropolitan canyons of Tokyo.

Traditional maps are not and will not be designed to provide the ultimate level of detail needed for some ITS applications, so accurate and reliable location information must be developed separately. For example, accurate detailed Digital Maps could be generated when white line painting takes place if the head of the white lining machine were to be equipped with a precision Digital Map logging system. However, testing would be needed to gain experience and verify results against the required performance before rollout across Europe. GIS (Geographic Information System) information, i.e. additional information such as lane location, maximum speed, types of curves ahead, etc., are not currently available. Collecting and providing this information would require a joint public/private effort. Additionally, there is also the problem of getting information on temporary changes, both in geometry as well as in the GIS type of information. For the many driver assistance systems an application independent solution to these issues must be pursued if further steps forward are to be made.

There is no consensus between the various stakeholders on the sorts of automated or assistance systems from which they can benefit. *Functional needs should be agreed on and, as only technologies that can be shared by different users seem to be feasible, win-win combinations of functions and technologies must be sought.* This holds for the regular operations as well as for all exceptional situations where infrastructure support is needed and for temporary changes where communications must be available.

Therefore, research is needed into:

- The need for ADAS supported functionalities by the various road operators all over Europe;
- The reliability issues and potential technologies of the various ADAS functions in all of Europe;
- The overall system concept or architecture encompassing the whole of ADAS to meet the liability and reliability issues;
- The deployment scheme backed by an EC policy for a coherent set of ADAS services.

If the above-recommended research is not undertaken, efficiency and reliability of most ADAS services will suffer; some may never get beyond demonstration stage. This will affect transport efficiency as well as Europe's industrial competitiveness with growing research programmes in the USA and Japan.

4.2 Co-operative Vehicle Highway Systems

Manufacturers are continuing to develop increasingly sophisticated vehicle-based systems to improve safety and support drivers. Vehicle technologies include those for lateral and longitudinal support and control (e.g. Adaptive Cruise Control and Lane Departure Warning) and those for driver comfort, convenience and safety more generally (e.g. navigation assistance, driver alertness monitoring, and journey time monitoring). Intelligent infrastructure technologies are related significantly to urban and interurban traffic management (e.g. UTC, access control, ramp metering), but geometric design standards take no account of the potential from new vehicle-based or co-operative opportunities. Both sets of technology can and will influence factors such as capacity, journey time, reliability and driver behaviour. Two forms of co-operation: vehicle-vehicle and vehicle-infrastructure have been shown to offer additional benefits to drivers and infrastructure operators (Bishop, 2001; MacNeille and Miller, 2004). Co-operation between vehicle and highway systems is necessary so as to:

(a) Maximise the opportunities of the various systems working together to meet industry's market-based financial objectives, and governments' broader social and economic objectives. (For example, some stop-and-go systems may be integrated with UTC systems to increase capacity and liberate road space.)

(b) Reduce the possibility of combinations of technologies working together and with drivers in such a way as to lead to situations where capacity or safety may be unnecessarily poorer than they might otherwise have been or expected. (For example, headways and associated driver behaviours with Adaptive Cruise Control systems may result in unnecessary delay if generally used with some ramp metering systems.)

4.2.1 Vision

In a fully co-operative vehicle/highway system, vehicle-based technologies will work actively with highway technologies to develop safe and efficient movements. A knowledge of vehicle locations relative to infrastructure and/or other vehicles and ways of communicating are crucial elements for a co-operative system.

Co-operative vehicle/highway systems offer the long-term prospect of vehicles driving autonomously, thereby removing the risks and inefficiencies resulting from driver performance limitations and driver-to-driver variabilities. Co-operative systems can operate at a series of levels, but the ultimate vision is for a driver to intervene only once at the start of a journey to identify the desired destination. The vehicle, in co-operation with highway systems, will then take the driver safely to the destination. Clearly, such a vision could only be achieved in the distant future, but there are a series of more achievable short-term scenarios which will contribute to safety and efficiency in particular types of location or operating conditions. What is clear is that the concept is exciting and has attractive potential benefits for all stakeholder groups: road users, government agencies, industry, service providers and road operators.

Co-operation can exist at three main levels:

(a) The vehicle receives information from the highway. In this situation the vehicle must have sufficient sensor technology to locate itself with respect to other road users and the highway information will enable the vehicle to 'anticipate' future conditions. (Such conditions may relate to static information such as road curvature or signal changes.) Magnetic or other markers may be used for more immediate location to enhance lane departure warning/lane keeping systems. This level of co-operation would substantially enhance the performance of vehicle-based systems. (Vehicle-to-vehicle communication would also enhance vehicle-based systems, but would suffer from inconsistent information depending on the locations of other vehicles at the times of need. Vehicle-to-vehicle communication would have separate benefits, particularly in enabling the actions of vehicles ahead in the traffic stream to be anticipated.)

(b) The highway receives information from the vehicle. The information may be used to supplement detector information on flows, queues, journey times, etc. to enable the highway operator to better manage the network through information and control technologies. Control may operate at a local level where knowledge of individual vehicle reactions may trigger immediate and short-term system responses to reduce accident risk or enhance capacity. (Advanced UTC systems such as SCOOT and UTOPIA make such changes at present, but detection is limited to specific measurements at certain locations only.)

(c) Fully co-operative systems where vehicle and highway information is exchanged for general benefit. For example, this approach would enable the headways of vehicles approaching an interchange to be manipulated to match metered entry flows to achieve and maintain optimum capacity, i.e. intelligent merging.

Other functions which vehicle/highway co-operation could perform include intelligent access control where, for example, vehicles meeting certain emissions standards could be given priority access.

A general outline of how development may progress is shown in Figure 6.

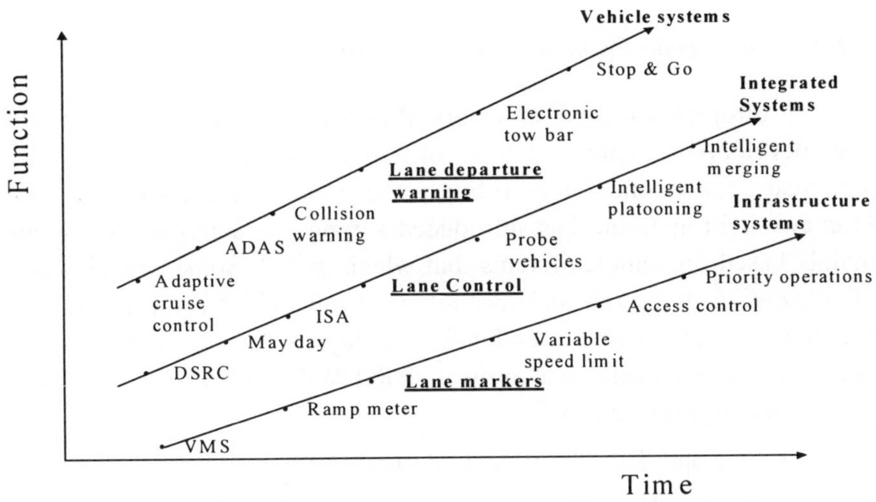

Figure 6: Development of Vehicle and Highway Systems

– Vehicle Systems: Each individual technology (e.g. lane departure warning) is being developed by the vehicle industry in isolation.

– Integrated Systems: Require the collaboration of vehicle manufacturers and infrastructure providers.
– Infrastructure Systems: Technologies developed by highways agencies and governments in isolation.

4.2.2 *Background*

A considerable amount of relevant ITS research has been undertaken, particularly in Europe, North America and Japan. However, until recently, the research has generally been concerned separately with either vehicle-based technologies or highway-based technologies, because of the different objectives, funding and decision processes, and timescales of the stakeholder groups involved.

4.2.2.1 European Framework Project results

Projects in the Framework Programmes since the late 1980s have developed a number of relevant new ITS technologies for vehicle and road applications. More recently the e-Safety Initiative in the 6th Framework Programme has introduced a range of activities which are mainly based on vehicle systems, but which include some consideration of integration. Research and technology development projects are also focused on safety systems and info-mobility systems or infrastructure management, with only one project, HIGHWAY, directly overlapping vehicle and highway systems.

Although Europe has developed a substantial knowledge base and capability of a wide range of systems in practically all areas of ITS, the ROSETTA project found that the focused development of individual applications from initial research to pilots led to a situation where subsequent synergy was not possible. The development processes were often technology led. This is particularly identified in a lack of integration between road infrastructure and in-vehicle systems. These understandings helped to define 6th FP directions.

Understanding the levels of accuracy needed for a wide range of transport applications and developing clear approaches to their achievement is crucial to integrated systems. As part of the studies into navigation systems carried out by ROSETTA a comprehensive table of accuracy and reliability requirements has been developed (ROSETTA, 2004 (9)) and discussed in section 7.2.

The objective of the Intelligent Vehicle Initiative (IVI) in the USA (Hanowski et al., 2002), and more recently in the Vehicle Infrastructure Integration (VII) (U.S. Department of Transportation, 2005), has been to advance the safety, efficiency and security of the surface transportation system, provide increased access to transportation services and reduce fuel consumption and environmental impact.

The USA IVI programme has achieved a high level of success, with federal targets generally reached. The work has been done with collaborative industrial partners from the USA and Europe. The success can be attributed to the vision and willingness of the policymakers and the open attitude of the suppliers. The more recently announced VII programme is beginning to gain momentum and is based on creating an enabling communications infrastructure which will support a spectrum of public authority and vehicle manufacturer interests. These range from signal pre-emption/electronic charging to remote diagnostics. Considerable emphasis is being placed on addressing the technology foundation of spatial representation, positioning technologies/ accuracies/communications latency, strategic development partnerships and developing leverage of the social and commercial potentials. A large number of vehicle/highway applications have been identified and business cases are being considered.

4.2.2.2 Traditional infrastructure systems

Technology has been developed over several decades to monitor traffic and improve network performance through information and control systems. Detector technology is continuing to advance with more recent emphasis on new functionalities, such as pedestrian detection and vehicle identification, and new technology applications, such as video based detection. Knowledge of network conditions is increasingly used to manage traffic more efficiently through the use of traffic signals (e.g. ramp metering and urban traffic control). In general, infrastructure systems are policy driven, respond to demand as it occurs on the street, and take no account of any intelligence which may be in the vehicle.

4.2.3 *State-of-the-art*

The following are a few selected examples of ITS functions in use today or at a very late stage of development; they represent the state-of-the-art of infrastructure-related systems with interactive options.

4.2.3.1 *Individual automated passenger transport – people movers*

Short distance transport systems can act as feeders for public transport hubs. They can also make the use of vast airport parking lots more attractive. A niche market in the transport chain is that of the fully automated people mover. For a long time warehouses have used pick-up robots and automated container transporters have been used at container depots and ports. Less known is the use of similar technology for automated vehicles in amusement parks. This technology has been operational for more than a few years for the transport of people; the systems have shown to be a good solution for shorter distances. Their routes are reprogrammable; their operations are accepted and appreciated by the public.

In general the possibilities these system could offer are little recognised by road operators - perhaps because most of these systems come in the form of fixed rail systems. However, free ranging versions, that only need a magnet in the road as position confirmation, can be used as regular vehicles on regular roads.

The Rotterdam-Capelle automated transport system uses individual battery driven vehicles to take passengers from a public transport hub to a central business district zone of office buildings (Parent and Gallais, 2003). The users of the automated transport system like the experience;

research shows that the passengers are extremely satisfied with the service provided. The network has now been expanded, with the number of vehicles as well as their carrying capacity increased. The vehicles, known as 'hoppers', automatically charge their batteries at a loading station. The infrastructure as well as the programming is similar to that used in the automated container terminal in the Port of Rotterdam or for all sorts of transport in Tokyo Disneyland.

4.2.3.2 *Road curve ahead warning*

When driving on an unfamiliar road it is easy to go into a bend at an excessive speed. A road curve warning system would advise the driver of the safe speed to enter the curve ahead in time for him to reduce speed to that level. The road curve ahead warning system requires a digital map and a GPS system or a roadside system where GPS reception is poor. Many demonstrations, such as during PROMETHEUS around Gothenburg Sweden (Augello, 1991), have shown that beacons along the road could be used to give timely warnings. This is again an example of a function for which various infrastructure-based solutions are possible.

4.2.3.3 *Route Guidance*

The in-vehicle route guidance systems available today use GPS and a digital map. The very first major field trial was the LISB system in Berlin (Sparmann, 1991) where infrared beacons were used to communicate with vehicles and a central system was calculating the preferred route. A few other pilots in which routes are calculated centrally followed later on. They showed an interesting advantage over the vehicle-based approach. Not only does the traffic management authority get a better insight to the origin-destination patterns, but since the service is not restricted to drivers, travellers using handheld units can also get advice for trips by public transport as well as on foot.

All navigation systems require accurate location, but they can all be served by one common system which must therefore be the most accurate. More than one location system can be provided for full coverage in difficult regions. Governments and industry should join forces to provide a combined map and location system.

4.2.3.4 *Source information*

Road authorities are responsible for providing good quality marking on the roads. Painting new lines could easily be combined with the recording of the GPS coordinates of the nozzle. Such arrangements would gradually produce perfect map source information and is an ITS version of providing maps.

4.2.4 Issues

The potentially large gains in safety and efficiency which the more advanced co-operative vehicle/highway systems could achieve will require a step change in the way in which industry and governments work together. It is difficult to determine an evolutionary approach, or one in which revenue/benefits are generated in the short term against substantial early investments. The US approach has been to involve technologies which have a wider market appeal; the beacon based technologies may not of themselves deliver all the potential co-operative vehicle/highway benefits, but market drivers may include banking and a wide range of financial and other transaction opportunities. The current CVHS study in the UK is focused on the development of outline business cases (Crawford, 2003).

Many of the benefits of co-operative systems will accrue to the society more generally as economic benefits from accident and congestion reduction. At present, there is no mechanism to bring together the economic benefits to governments with the market driven financial benefits to industry. This may be best addressed by trying to understand what business cases are required by the various stakeholders, and to identify the research needed to populate these business cases to enable decisions to be made. Inevitably, whatever the scale and quality of the research inputs, decisions will require more general commitment to the vision, but this process is poorly understood. The RDS/TMC system which has been implemented across Europe is a much simpler system/concept, yet took considerable time to negotiate.

Research issues for which there are no currently agreed answers include:

- How far can in-vehicle systems, designed to give the driver comfort and improved safety, be influenced to support the broader objectives of overall efficiency and safety?
- How can infrastructure-based systems be made more efficient by understanding the needs and capabilities of the driver with an understanding of in-vehicle systems?
- How can the promised efficiency, safety and environmental benefits of combined in-vehicle and infrastructure systems, including vehicle-to-vehicle communications, be realised?
- How best do we define common objectives, evaluation procedures and development path?
- To what extent can a combination of integrated in-vehicle and out of vehicle systems lead to modified driver behaviour more generally?
- How will the new satellite systems affect the quality and availability of GPS location and what are the long-term development pathways for vehicle communications?
- What facilities and special requirements are needed at interchanges to allow for multimodal operations and access control to motorways?
- How important is European collaboration?

4.2.5 Future opportunities

Recommended research actions are as follows:

- Research to validate benefits suggested by early simulation studies concerning the inter-relationship between improved driver comfort and increased network efficiency;

- Research towards understanding the impacts of ADAS on drivers and driver behaviour and how this can be adapted to a beneficial network model;
- Develop a long-term strategy with agreed goals to merge the independent development of systems to support the driver and infrastructure systems required by authorities;
- Research into the new assessment methodologies required to understand the true cost-benefits of driver assistance devices that are likely to change the way we drive;
- Research into the behavioural changes from external speed enforcement combined with in-vehicle systems, which may have a wide impact on safety, capacity and the environment;
- Research to help to identify the most suitable technologies and systems architecture for vehicle-vehicle and vehicle-roadside communication for direct vehicle control applications;
- Research to assist network operators to realise interchanges that allow for freight, public transport and private transport operations, interacting directly with the motorway network;
- Research into standardisation.

EU collaboration will be key to the successful implementation of most if not all of these activities.

4.3 Human Machine Interaction

4.3.1 Vision

Intelligent Transport Systems have the potential to save thousands of lives on European roads as well as improving our environment and economic performance while also making travel more efficient and enjoyable. In this vision, technology can be used to support drivers and other travellers in all their pre-trip and within-trip activities through straightforward, safe and effective processes of interactions. However, ITS can also be part of the problem if poorly designed, so human factors and the design of human-machine interaction is critical in ensuring that there are not new hazards for drivers and other road users.

To support the vision of the benefits that can be achieved through ITS three strategic aspects need to be addressed:

(a) *Ensuring that information and communication systems do not create new safety problems*. Essentially, the issue is concerned with driver distraction and overload. The use of in-vehicle equipment, including mobile communications devices, PCs, e-mail and Internet, needs to be carefully studied. Widespread deployment should not take place until the problems of poor human-machine interaction have been solved. One specific measure is to develop the EU 'Statement of Principles on HMI' to include an assessment process to ensure that only those systems complying with the principles are installed within vehicles. In parallel, users (and this includes employers and hire companies as well as drivers themselves) of both installed and 'nomadic' devices need to be better informed concerning the risks and suitable enforcement measures developed. Work is underway through the eSafety initiative but there is a dearth of underpinning research to validate the link between human and system characteristics and safety.

(b) *Ensuring that Driver Assistance Systems (DAS) with the largest expected benefits are given greatest prominence*. Assistance systems that automate some aspect of the driving task have the greatest potential (if properly designed) to optimise the driver's interaction with the road environment and hence lead to dramatic road safety improvements. Ongoing research and evaluation of early systems is needed to identify and validate expected safety improvements. Examples of systems expected to offer particular benefit include those that help the driver manage speed (intelligent speed adaptation), ensure seat belt wearing and avoid impaired driving (e.g. caused by alcohol, drugs or fatigue). Such systems need to be thoroughly researched and promoted including technical issues such as robustness and fail-safe modes, and human issues of acceptability and behavioural adaptation.

(c) *Ensuring that driver assistance systems coming to the market do not create new safety problems.* Ideally, commercial DAS are matched to user needs and expectations, but this is not completely possible because of technical and economic factors and any mismatch raises potential safety issues. Research is therefore needed to understand these human issues (mental models and expectations) and to develop mitigation strategies such as driver information and education.

4.3.2 State-of-the-art

4.3.2.1 In-vehicle ergonomics and distraction

Advances in vehicle instrumentation are driven by customer demand, the desire of manufacturers to add value, and by technology which offers increasingly sophisticated and cost-effective options. According to ARC Group's Automotive & Freight Telematics Strategic Report (ARC Website) the world market for in-vehicle telematics systems was forecast to grow to more than 50 million units by 2005. Emerging new services include location-based and navigation services, logistics and fleet management, information services and office applications. Multimedia and entertainment services are predicted to be among the highest value applications.

In-vehicle information and communication systems can assist, for example, with pre-journey and on-trip planning, bringing benefits in terms of better use of the road network. They can reduce uncertainty and stress, thus calming drivers and potentially contributing to safety. However, poorly designed systems could adversely affect driver behaviour, and hence safety, by distracting attention from the driving task. Also, if they supply inaccurate, untimely or misleading information they could prompt a driver to take inappropriate action, thus endangering themselves or other road users.

There are a number of 'vehicles' that are used to contain information about HMI design and assessment, distilled from research studies and other co-operative agreements. These include published material such as papers and books, international standards, Codes of Practice, type approval and other laws. The style and content vary but in general one of three approaches is adopted:

- Design approach – e.g. manuals, text books;
- Process of development or assessment;
- Performance measurement (performance of the system or of the user or both).

4.3.2.2 Recent and current EU research

EC projects, since the beginning of the Framework programmes, have been tackling HMI as a 'horizontal' activity with projects such as STAMMI, HARDIE, HOPES, EMMIS, and GEM. There have also been projects concentrating on the special HMI needs of elderly and disabled drivers (e.g. TELAID and EDDIT) and projects aimed at detecting driver impairment such as SAVE. More recently, HMI activities have been undertaken within individual projects such as UDC, LACOS, ITS-WAP and COMMUNICAR. From all this work, a better appreciation has been gained of the advantages and disadvantages of HMI assessment techniques and of the importance of driver attitudes and behaviour.

In the 5th Framework Programme the European Commission and the different project partners have spent over 250 million Euros on more than 80 research projects that are related to the ROSETTA work areas. Specifically, work has been undertaken in the field of Human-Machine Interface and an overview of relevant 5th FP projects is shown in Table 5. The presented areas are derived from Deliverable 6 of the ROSETTA project.

Table 5: Relevant 5th FP research activities

Area	Project
Telematics in public transport	ITSWAP
In-vehicle ergonomics and distraction	HASTE
	COMUNICAR
Driver assistance systems	ADVISORS
	STARDUST
	EDEL
	EUCLIDE
Influencing behaviour	ROADSENSE

4.3.2.3 Standards

Standards are 'Documented agreements containing technical specifications or other precise criteria to be used consistently as rules, guidelines or definitions of characteristics, to ensure that materials, products, processes and services are fit for their purpose.'

HMI standards for ITS products are being developed in parallel with the technology and market introduction. They involve the interaction of drivers with in-vehicle equipment and therefore are rather different from those concerning, for example, communication protocols or databases.

There are three main approaches to standardisation; procedure, design and performance. Procedural standards concern the process of doing something and may say little about what is actually done within the process. Examples are ISO 9000 and ISO TS16949. These concern quality and processes to deliver consistent performance and organisations can be externally audited for compliance. Design standards specify principles and features appropriate to a product. There may be options within it or minimum criteria. An example is the gearshift patterns for automatic gearbox passenger vehicles. Performance standards require the specification of performance of equipment or of users in relation to use of the equipment.

Performance standards are favoured by manufacturers as they can specify 'what' has to be achieved (to comply with the standard) without specifying exactly 'how' it has to be brought about. In the area of interaction with in-vehicle equipment, the research base on which to agree human performance criteria is not well established.

4.3.2.4 HMI principles and guidelines

European statement of principles

In December 1999 the European Commission adopted the European 'Statement of Principles' (ESoP) in acknowledgement of the importance of the Human Machine Interaction (HMI) for in-vehicle telematics. The principles apply to information and communication systems that are intended for use while driving. The principles apply to these systems whether they are directly related to the driving task or not. They also apply to both portable and permanently installed systems, both original equipment manufacturers and after market systems, and for all road vehicle types.

The EC principles cover aspects of overall system design, installation, information presentation, interaction with displays and controls, system behaviour and information provided about the system. The interactions between multiple in-vehicle devices and issues relating to multiple in-vehicle systems such as consistency and compatibility are not covered by the ESoP; nor are speech input and output interfaces.

The European motor manufacturing industry has formally declared its intention to follow the principles and Member States have been invited by the EC to study its impacts.

AAM American statement of principles

In July 2000, the National Highway Traffic Safety Administration (NHTSA) held a public meeting to address growing concern over motor vehicle crashes and driver use of cellular telephones. During this meeting the industry was challenged by NHTSA to respond to this rising concern. As a consequence, the Alliance of Automobile Manufacturers are developing a 'best practices' document to address safety aspects of interactions with in-vehicle information and communication systems and the AAM American Statement of Principles was developed based on the ESoP. Members of the Alliance of Automobile Manufacturers (AAM) made a formal commitment to design and test future telematics devices in accordance with the guidelines.

The most recently published guidelines (2002) include some performance criteria and verification procedures. Although many of the AAM guidelines are based on well-documented installation and location principles, the information presentation guidelines have been criticised for being elusive as they involve quantifying driver behaviour and performance. The use of the AAM Statement of Principles by system manufacturers and designers in the US has not yet been assessed as development of the guidelines and assessment procedures are ongoing.

Comparison of EC SoP and AAM guidelines or principles

The AAM and EC Statement of Principles (SoP) do not differ greatly, although some changes, additions and exclusions have been made for the AAM. The main difference between the EC SoP and the AAM SoP is that the AAM principles take a further step by providing criteria and verification procedures by which to assess a system on most principles.

A further difference is that the AAM excludes audible vocal interfaces whereas the EC SoP includes such systems.

The AAM SoP also words some guidelines differently. In some cases the wording is changed slightly for clarity. In many cases the same guideline as presented in the EC SoP has additional explanatory text, again to clarify issues or include further issues. For example, additions are made in the section considering hands-free equipment. AAM additions are included for push-to-talk systems and to address concerns related to hands-free headsets, such as drivers donning headsets whilst driving.

Japanese guidelines

The Japanese Automobile Manufacturers Association (JAMA) produced a set of guidelines for In-vehicle Display Systems in 2000. The document includes guidelines on display location, display requirements whilst the vehicle is in motion, operational requirements whilst the vehicle is in motion and software provided by third parties.

More recently the Ministry of Land, Infrastructure and Transport organised a committee in 2001 to make a regulation to ensure the safety of navigation systems. The result of this was the production of general HMI guidelines which refer to the JAMA guidelines and the European Statement of Principles. The quantitative guideline proposed for 'Interaction of Display and Operation' describes the display and control locations, the display requirements while the vehicle is in motion and operational requirements while the vehicle is in motion. It also includes a specific method to assess visual distraction.

The National Police Agency (NPA) has established a guideline that no television or e-mail facility should be operational and accessible to the driver whilst the vehicle is in motion. Further, the National Police Agency and the Ministry of Land, Infrastructure and Transport organised a Traffic Information Consortium (TIC) to discuss the provision of high quality information, the promotion of the participation of private companies and the development of new technologies. The TIC proposal relates to safety issues, proposing that rules are necessary for private companies to ensure traffic safety. For example, keeping information accurate, prohibiting unsafe navigation and ensuring safe HMI and information.

4.3.2.5 eSafety

The eSafety Forum was established by the Commission (DG Information Society) in close collaboration with the industry, industrial associations and public sector stakeholders to look at both safety and market issues in the implementation of driver information and assistance systems as a contribution to European road safety improvement targets. In November 2002 the final report of the eSafety working group on road safety was published containing 28 recommendations (European Commission, 2002).

The eSafety steering group established a working group on HMI to tackle the important issue of driver interaction with on-board devices such that HMI does not become a barrier to deployment.

The HMI working group presented its draft recommendations at a Brussels workshop in June 2004 (eSafety website). In outline, these involved:

- New Regulations for nomadic devices concerning physical construction and secure vehicle fixing;
- Increased co-operation between vehicle and system manufacturers;
- Service providers to develop 'Safety Agreement' concerning presented information;
- Fleet owners and employers to be reminded of their responsibilities for health and safety;
- Authorities responsible for dissemination of HMI guidelines, consumer information, monitoring and enforcement;
- Authorities to seek self-commitment of nomadic providers to the European Statement of Principles on HMI (ESoP).

Concerning the principles themselves, it was noted that the EC will write a communication on the ESoP and HMI at the end of 2004, and three additional recommendations were made by the Working Group:

- Develop ESoP to include 'Post-Manufacture' issues of use by employers, hire companies and drivers;
- Explore commonality with the AAM guidelines in the United States and the Japanese approaches;
- Better dissemination and monitoring of ESoP is required.

4.3.2.6 Mobile telephones

Most countries currently deal with legislation for mobile phones separately from other In-Vehicle Information Systems (IVIS). However, agreement on common legislation for mobile phones has not been reached. Different approaches to legislation of mobile phones whilst driving exist, both internationally and even within the Member States of the EC and the US. In some countries and states the hand-held mobile phone is banned. In other countries both hand-held and hands-free mobile phones are banned. In some countries mobile phones are not banned whilst driving, but are considered to be covered by dangerous driving laws. Finally, in some countries mobile phone use whilst driving is not addressed at all by legislation.

One view is that it is a reasonable restriction to ban hand-held mobile phones; another is that such a ban does not address cognitive distraction. A further argument is that singling out specific technologies for regulation is not the correct approach and that mobile phones should be covered by the same legislation as that provided for all other IVIS.

This is a dynamic area and one where interpretation of the exact situation is difficult or controversial in some countries (ROSPA, 2002).

4.3.2.7 Assessment methods and assessment variables

Assessment methods use environments within which data is captured. Example environments include mathematical simulations, driving simulators and test tracks. Assessment methods also use tools to obtain data. Example tools include video recorders, eye trackers and questionnaires and the data generated using tools is processed to yield variables.

The principal relevant characteristics of assessment methods and the variables produced are:

– Validity – extent to which the variable is diagnostic for the concept being investigated;
– Reliability – reproducibility of measurements over time;
– Sensitivity – ability to measure small changes in a variable.

In addition to validity, reliability and sensitivity the choice of an assessment method will be influenced by other practical factors including the cost and availability of environments and tools, and the time and effort required for data gathering and processing. Methods for assessing the performance and safety of a system's HMI may be categorised into three levels as in Figure 7.

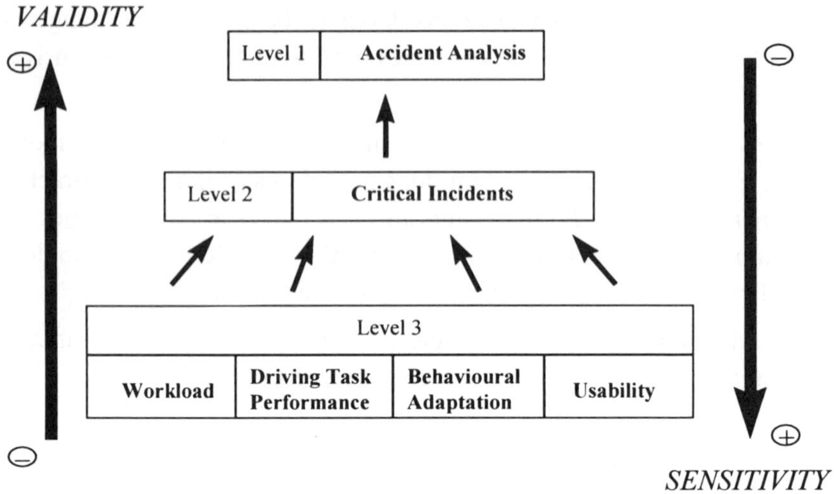

Figure 7: Three-level assessment method categorisation

Moving from level 1 to level 3 is associated with an increase in sensitivity but a decrease in validity.

4.3.2.8 General points about assessment

There appears to be a lot of knowledge about how to undertake human factors assessment, but there is also a wide variation in what is done and how it is reported. Some commonality would be extremely useful and the following key points provide basic guidance:

- Evaluations should be done with Human Factors experts and IVIS end users. The end users should be experienced drivers who are familiar with the system;
- Both male and female drivers and the young and elderly should be tested. A minimum of ten people should be tested;
- Preliminary desktop and driving simulator evaluations are often necessary to identify serious safety and usability concerns before systems can be tested in real situations;
- Checklists are a useful preliminary tool for expert assessment.
- Full evaluation ultimately requires road tests;

- Comparisons should be made in driving behaviour with and without a new system. Any significant degradation in performance when driving with the system should be a concern.
- A useful categorisation of test methods is:
  - Measure the driver;
  - Measure the vehicle;
  - Measure interaction with the system.

4.3.2.9 Driver assistance systems

In-vehicle systems, such as Adaptive Cruise Control (ACC) and collision warning are described by manufacturers as providing ‘driver assistance’, the implication being that the driver remains responsible for safely manoeuvring the vehicle.

The implicit assumption is that the driver is part of the road/vehicle system and will interact in certain ways in response to the road environment and to the technical system. However, unlike quality assured production items, the variability encountered is quite large.

For driver assistance systems to be used efficiently and effectively, certain basic human factors requirements need to be achieved. These include reliability over time and with regard to external influences, robustness in case of a system malfunction or misuse, perceptibility of the human-machine interface, comprehensibility and predictability of system functionality, and controllability in all situations.

In addition, there needs to be consideration of ‘foreseeable misuse’. This recognises that drivers are human and do not always behave as instructed or expected. With regard to traffic safety it is important to know what risk the individual driver is willing to accept. Underwood et al. (1993) consider the compensatory process that takes place in reaction to the introduction of safety measures. Using drivers’ motivations (other than safety), as well as expected accident cost, their model determines the expected benefit of a safety measure. Based on the assumption that the aim of road users in making a trip is to maximise the benefit of the action, risk compensation is posited to occur as road users respond to changes in the system. This response ultimately ensures that their personal needs are achieved.

It is therefore necessary to examine the extent to which the systems are assumed to be 'beneficial to safety' and the probability of risk compensation as reflected in risky behaviour. Apart from an increase in dangerous driving manoeuvres, such behaviour can also be reflected in 'testing of the limits' or 'risk seeking'. One other important aspect of misuse is the intentional use of a system beyond its known system boundaries. In general, it should be noted that risk and misuse potential of driver assistance systems should always be limited to a minimum by design.

4.3.3 Issues

There appears to be a consensus that distraction is an important issue relating to in-vehicle systems. However, there is little legislation concerning IVIS for most countries, although there are 'guidelines' contained in the Statement of Principles. The effectiveness of the guidelines has only been assessed in a few countries and recommendations given for future developments differ between those countries. There is a consensus that more research and development is needed to produce better criteria for the principles and that there is a need for international co-ordination and agreement.

One exception to this is the area of mobile phones where agreement on the extent of cognitive distraction does not seem to have been reached. Different approaches to legislation of mobile phones whilst driving exist, both internationally and even within the Member States of the EC and the US, as noted earlier.

Clearly there would be a benefit in having a common approach to mobile phones as with IVIS. This would be particularly beneficial for users as this would dispel confusion when travelling across state borders or into different countries.

The question remains as to whether hands-free mobile phones are really better than hand-held. A number of research studies have found the cognitive distraction imposed by mobile phone conversations to degrade driving performance and further research is needed in this area.

Debate exists as to what the best approach is to take in terms of guidelines and legislation for IVIS. There is a view that all IVIS should be legislated for together, therefore mobile phone legislation should not be separate. However, it is difficult not to be technology specific as different issues are related to each separate technology.

4.3.4 Future opportunities

Understanding user needs and the process of diffusion of new systems into the market is vital. Focus groups, etc. have a part to play, but future users may not have a good perception of their needs for new services. Future needs and wants have to be discerned and implied from developments in society. In particular, it should be noted that different user groups will have different needs.

Manufacturers are very active in developing and marketing ITS/ telematics applications that will provide enormous increases in the functionality available to drivers. There is great excitement concerning how these systems might enhance the driving experience but safety and human factors efforts lag behind electronics development. In particular, driver distraction needs to be tackled if accidents and a tarnished public image of certain products are to be avoided.

In the area of interaction with in-vehicle equipment (both driver information and driver assistance) the research base on which to develop widely agreed performance standards is not well established. However, it is likely to be possible to agree a number of general rules. In the long term, the most effective means of minimising risk to drivers will be through improved product design.

Development of general design principles and Codes of Practice covering the design process are probably more effective and flexible ways of promoting good HMI design than through standards and legislation (although the different approaches all have strengths and weaknesses).

It is relatively easy to draw up lists of 'research needs' in HMI but different organisations and actors will have different perspectives on the priorities. Particular areas to highlight could be standardisation of testing methods, better information on 'normal' driving and research into workload managers to avoid driver overload. Clearly, there is much work to be done.

Practical assessments of existing products often identify systems and situations where insufficient attention has been paid to the variability in human performance in the design or use of the HMI. Key stakeholders, therefore, still need to be better informed concerning HMI and safe in-vehicle interactions. More work needs to be done to engage and influence manufacturers and employers, as well as drivers.

4.4 Emergency Response

4.4.1 *Vision*

Whatever the safety impacts of ITS, there will always be road traffic accidents resulting from human or technology failings. In the most severe crashes, people may be killed or seriously injured and rapid medical response can substantially increase the likelihood of survival and reduce the long term effects of injury. The so-called 'Golden Hour' rule (confirmed by studies made in the STORM project in Germany) claims that the lives of between 20-40 % of seriously injured people can be saved if they receive hospital care within 60 minutes of an accident. The probability of survival is further increased if first aid is given at the site of the accident before transport to hospital (in the 'Golden Ten Minutes'). Recent analyses in Europe have estimated that the use of telematics can reduce the reaction and intervention time to emergency calls by as much as 30 %, and that emergency calls generated automatically from vehicles can increase the probability of survival in the case of road accidents by 15 % (ERTICO, 1997).

The vision for emergency response is that the location of an accident and an indication of severity is automatically sent to the emergency services so as to trigger a response for a more timely and effective treatment of casualties. In addition to medical services, quicker response by police and breakdown services will generally lead to reduced delays to other traffic. Also, such systems may be personal rather than vehicle based and be used to develop a wider range of medical and related response services.

4.4.2 *State-of-the-art*

Over the last decade, a series of research projects and field trials carried out in both the United States and Europe have provided good insight into the fundamental technical, legal and organisational requirements of emergency call support for road users. In Europe, a legislative basis has now been laid down, and efforts are being made towards the achievement of a harmonised system which is operable in all Member States.

The potential of ITS applications in the management of emergency calls was first raised in the 1990s in the USA. The benefit of automating information flows connected with road accidents was brought to the attention of the ITS world at the National Conference for Rural IVHS at Keystone, Colorado in February 1993, and triggered a series of field tests.

4.4.2.1 *Research carried out in the United States*

One of the first operational tests was the Colorado Mayday project, led by the Colorado DOT (Department of Transportation). Its aim was to:

- Create an organisational infrastructure that would co-ordinate the activities of the many agencies, public and private, that need to co-operate in an emergency response network;
- Define a standard communication link between the vehicle and the control centre;
- Reduce the overall cost of the system to a level low enough for the average consumer.

The last goal, especially, led to useful insight into the way GPS information could be used. This project was concluded in 1998.

The Washington State 'Puget Sound Help Me' (PuSHMe) project tested the possibilities of using pager and cellular phone technologies in combination with GPS. Systems with and without voice link were tested. This project finished in 1999.

The Automated Collision Notification Project was a New York State Mayday project that focused on the technical capability. It integrated crash sensors to help trauma teams assess the required Emergency Medical Services (EMS).

A Minnesota-based project, Mayday Plus, which finished in 2000, tested a state-wide emergency response infrastructure, as well as resolving jurisdictional issues and concluded that the processing of emergency calls can be extremely complicated.

Between 1995 and 1999, a Multi-Jurisdictional Mayday (MJM) group was active as a forum for critical analysis and information exchange on the risks, barriers and opportunities associated with Mayday deployment. It represented the needs of public and private sector response agencies in the Mayday arena, with particular emphasis on standards and systems' functional requirements.

One of the important conclusions of the field trials carried out in the USA was that the many different types of Mayday situation could all benefit from improved notification systems, and in particular an increase in the speed with which information is relayed to the emergency services. Further analysis led to the suggestion that the architecture for dealing with road accident emergencies could be very similar to that for emergencies involving the transport of hazardous goods. For inland waterborne traffic, such notification arrangements have already been developed in Europe, and joint future development is foreseen.

Extensive field trials in the USA showed also that a direct video link between the EMS crew in the field and the Trauma Centre could improve the quality of response. Satellite connections were used in these trials, but this is a costly solution. Two alternatives emerged from these trials. Simple Polaroid pictures of the victims still in the crashed position can give a trauma team valuable information about possible injuries, but the latest mobile phones with cameras offer a simple and cheap opportunity to give a Trauma Centre the crash pictures it needs. Here technology development came just in time with a good solution.

In line with the approach advocated by, amongst others, the European Commission and USDOT, the aim in the United States has been to develop an open architecture able to deal with emergencies relating to a wide range of different situations, including radiation accidents, chemical and oil spillage, forest fires, mountain and cliff rescue, maritime search and rescue, medical help for the elderly, and rail incidents as well as road accidents. The emergency notification procedure for all of these could therefore have a common basic architecture, but with additional 'information blocks' for each specific type of application.

A further conclusion was that the procedures adopted for dealing with road accidents must take into account the implications of different types of call, i.e. calls for help made both manually (via fixed or cell phones) and automatically (generated from the vehicle), Good Samaritan calls made by third persons and providing 'secondary' notification of an incident, as well as calls made to private service operators for roadside assistance.

Overall:

- The system for the management of emergency calls regarding road accidents should be part of a more general system covering many different types of emergency;
- The underlying architecture should be open and flexible;
- In developing such a system, account must be taken of calls made both manually and automatically, as well as Good Samaritan calls and service calls.

4.4.2.2 *Research and consultations carried out in Europe*

In Europe, the process which led to the definition of the legal aspects of the emergency calls service began in 1991 with Decision 91/396/EEC which established that a single emergency call number, 112, should be adopted throughout Europe. This was followed up with Directive 98/10/EC of 26 February 1998, requiring Member States to put this number into operation.

In subsequent years a series of European projects, studies and other initiatives have systematically examined the issues involved in automating the process of making and processing emergency calls from vehicles. They have subsequently made recommendations to the European Commission regarding the various organisational, technical and regulatory aspects involved and in some cases also carried out pilot trials.

CGALIES (Co-ordinating Group for Access to Location Information by Emergency Services) was an EC-initiated IST programme (2000-2002) which included representatives of the telecoms industry and sought to reach consensus before the legislation became effective. It carried out a detailed examination of the technologies available for the location of the origin of 112 calls, and made recommendations regarding policies for the application and implementation of the basic infrastructure of the e-112 system ('Enhanced 112').

The EC 5th Framework Programme E-MERGE project, part of the INFSO programme, ran from 2002-2004 and had the aim of harmonising solutions at the European level for the management of emergency calls made automatically from road vehicles equipped with onboard devices. It sought in particular to resolve problems relating to emergency calls made in border regions.

The IST project, AIDER (2001-2004), worked on the development of on-board equipment providing automatic incident detection and communication with emergency centres.

A further European initiative, eSafety, was launched in 2002 to promote the development of integrated 'intelligent systems' for road safety. A discussion forum and work groups were set up to promote and monitor the creation and implementation of such systems. A Steering Group was instituted with the task of producing a functional European model for the management of emergency calls by the end of 2003. This focused on vehicle-based applications, but also considered other related developments (and hence also non-vehicle applications using the same architecture).

The aim of the LOCUS project is to provide the EC with support and expertise regarding the definition of a Location Based Emergency Service in Europe, taking into account the user needs, institutional issues, technology issues, markets and need for convergence with other applications.

A 6th FP IST project, RESCUE, a subproject of the Integrated Project, GST, which began in 2004, is focusing on the problem of automatically assessing the type of emergency involved, and forwarding this information to trauma centres and the emergency vehicles involved. A routeing system will then ensure that other road users are warned of the approach of an emergency vehicle.

In successive stages, these projects have been able to contribute, and are still contributing, to the definition of an architecture which will permit the complete end-to-end management of emergency calls in Europe for both vehicle and non-vehicle applications.

4.4.2.3 Requirements for an e-112 System

Ideally any e-112 system should be capable of the following services:

– *Provide accurate location information:*

  All emergency services and others who need to reach the site of an accident must have sufficiently accurate location data to know unambiguously where to go and also the best way of getting there, since routeing information which takes into account real-time traffic conditions is available.

– *Distinguish different types of mayday call:*

  When an automated 112 emergency call is received by the PSAP (Public Service Answering Point), it should be able to distinguish immediately between the different types of system, e.g. whether it comes from a private car, a vehicle carrying hazardous materials, or is a personal Mayday service call. It is also possible to recognise Good Samaritan calls.

Not only will e-112 offer services to drivers, similar systems will be available to monitor hazardous goods transport, hikers in distress, to serve maritime transport and also to support elderly people needing to get in touch with their 'home base'.

– *Know whether, and what kind of, medical assistance is required:*

  This is possible either through direct voice contact with a medical expert at the PSAP, or via a conference call service set up with a trauma centre. In any case, the caller always has a voice link connection in the case of a 112 call.

– *Identify the caller's phone number or network:*

  This makes it possible for contact to be re-established with the caller when necessary. For cellular phones, this information will be available automatically, so the PSAP can forward it to the other emergency services involved.

– *Language support:*

  For calls involving foreign drivers, conference calls will be set up by the PSAP with a service provider to permit translation when there is a language problem.

For vehicle-based systems, the following additional information is also available:

– *Vehicle identification:*

  Information is available not only on the licence plate and country where the vehicle is registered, assisting in the identification of the persons involved, but also on the type and colour of the vehicle, thereby helping the rescue teams find the vehicle rapidly.

– *Number of persons involved:*

  This information is available in some cases from a special onboard passenger-counting device, a webcam or mobile device in the vehicle, and can help the trauma centres plan ambulance services effectively when no voice contact can be made.

– *Crash/impact data:*

  Information from in-vehicle devices provide details which help the field response staff and PSAP know what kind of service is required and type of injuries. It includes crash type and severity (indication of rollover and principle direction of force), axis of acceleration, time-histories for the entire crash, and final resting position of the vehicle.

4.4.2.4 *Handling of the calls*

E-112 services will form a part of the system that also offers general personalised services such as 'yellow page information', route guidance and car maintenance. These systems may be marketed as part of the car manufacturers' range of services. Non-emergency calls in such systems will be handled by call centres. In the case of an emergency call, the call will go directly to the PSAP like all calls from dedicated e-112 systems. The service provider may have a party-line listen-in facility.

The procedure for the management of manual or automatically generated emergency calls made from vehicles is as follows:

- The emergency call generated automatically by the onboard system (IVS) is sent to the PSAP through the voice channel e-112. It is composed of two elements: one voice and one data component, supplying the minimum set of data (MSD) required to respond to the call, both though the same voice channel.
- The mobile telephone operator, responsible for sending the call to the most appropriate PSAP, adds further information concerning the CLI (Caller Line Identification) and the location data generated by the telephone network.
- If the driver subscribes to a private service, a second message is sent from the vehicle including extra data sent to the Service Provider who interprets the contents and provides additional static data (e.g. details of the driver, medical data, contract information, etc.). In this case the PSAP can interrogate the

Service Provider to obtain this additional information and complete the background on the emergency.

- For a foreign vehicle, if the driver subscribes to a private service and requires language support, the PSAP can establish a 'conference call' between the driver and the Service Provider to permit the necessary translation.

Note that for hazardous goods transport the haulier might be the service centre contacted, while for elderly person support the home base may take that role.

An example of a system for emergency call support designed specifically for heavy trucks is the Volvo Action Service. A driver in difficulty calls the number on the Action Service card and is connected to one of two emergency-assistance centres (in Gent in Belgium or Rugby in England). These centres are open 24 hours a day, every day of the year and can handle most European languages. The operator notes down the necessary facts about the vehicle and its problem. The operator can call up the vehicle's entire technical specification on a computer terminal. The exact location of the vehicle is identified with the help of a stored sequence of the latest GPS fixes from the vehicle and marked on a sophisticated mapping system. Appropriate rescue action is then taken.

Another example of a specific use of e-112 is the service for forest workers in Sweden. Their e-112 system includes an emergency button on their left shoulder and an emergency call gets them in touch with the Swedish PSAP: SOS-Alarm.

4.4.2.5 European directives

Recommendations made to the European Commission by the various research projects and initiatives have led to the formulation of a series of directives which lay down a regulatory basis for emergency call management in Europe. The most important are the following:

- The Universal Service Directive, 2002/22/EC, requires mobile telephone service operators to make available information relating to the location of emergency callers to the authorities involved in the management of emergency services. To make this possible, an exception to Directive 2002/58/EC, protecting the privacy of electronic communications, was stipulated for organisations involved in the emergency management, including medical services and the fire brigade;

- Directive C(2003) 2657 of 25/07/2003, which recommends that this procedure be adopted for all calls made to the single European emergency number 112. It required telephone operators to automatically forward to the emergency service managers information on the location of any calls made to a PSAP. For a transitory period, such information may however be sent to the centre on request. A key aspect of this recommendation is that Member States are required to adopt a common interface, with a common flexible protocol that will allow adaptation to any new requirements dictated by future developments. It is proposed that the protocol should be that defined by ETSI on the basis of the specifications provided by OCG-EMTEL ('Ad-Hoc Group on Emergency Telecommunications').

Further Directives and standards are planned. Draft standards on location referencing are expected in 2006. The implementation of e-112 will then be evaluated and new legislation drafted if necessary. With regard to location referencing, a considerable amount of standardisation work is going on already, e.g. LIF, Agora, Nexmap, GTP, GSP rescue.

Research and pilot trials by European and national projects have enabled important progress to be made with regard to the technical problems and questions of procedure. For example, an important procedural step regards the definition (produced by the E-MERGE project) of the MSD - minimum set of data - required to be able to respond effectively to automatically generated emergency calls. This consists of the following information:

- 'When' the emergency occurred;
- 'Where' it occurred, the location of the vehicle on the GPS satellite network (position and direction of movement of the vehicle);

- 'Who is involved', a description of the vehicle from which the call was generated;
- 'Who can provide further information', through the identification of a Service Provider with whom the driver has a contract;
- 'What help is needed', the 'seriousness' of the emergency, and the type of call, i.e. whether automatic or manual and, in the former case, which sensors have generated it.

Note: it was expected that the information on the vehicle occupants' identity would be also useful for trauma centres, but since rapid 100 % reliable checks on automatically generated data are not possible (and as the centres cannot take the risk that the IDs are not correct), it is assumed that standard procedures will be adopted for dealing with unidentified patients, involving the use of photo identification.

4.4.3 *Issues*

4.4.3.1 *Location Referencing*

Accurate location reference information is required by all parties that need to be directed to the site of the caller. If precise location is not available immediately, it is helpful for the PSAP if a rough position can be derived, followed as rapidly as possible by accurate positioning.

Until now, the 'best technology' for achieving this has not yet been defined. Although latitude and longitude co-ordinates give good enough accuracy for most non-vehicle incidents, it is not sufficient for locating road accidents. On many highways, it is necessary to know which carriageway is involved, and this requires direction of travel information. For a vehicle which has crashed into the median barrier, accurate infrastructure-based location referencing without direction of travel is very difficult, if not impossible to use. Incorrect information leads to unacceptable extra access time for emergency vehicles.

The required information can easily be provided by the caller's phone if it was working when the accident happened. In this case, agreement is needed on how to code the information. This can be deduced from a sequence of GPS fixes when the caller is actually driving, but will not be the case when the caller has crashed and needs help. It must therefore be assumed that either information on the direction of travel is given by the caller, or is derived. This could be done automatically by adding direction information or by storing GPS fixes every minute and providing the actual position plus the latest stored fix. If this solution is chosen, again an agreement is needed on the format of the information.

In the case of on-board GPS location facilities, the unit must be operating continuously in order to avoid a long start-up delay when a call needs to be made. With mobile communication needing to be hands-free in more and more countries, this should not pose a problem. Furthermore tests have shown that a general solution was needed to solve the problem of accident black spots where insufficient GPS information was available. From the Mayday tests, the suggested solution is to use the last GPS fix and the direction of travel information.

For manual calls made via landlines, rough location given by the first digits of the caller's number will be sufficient, while for mobile calls, the cell-ID represent one possible acceptable technical solution. In remote areas, satellite telephone communications are still the only way to communicate with a vehicle.

For landlines, precise location can be derived from the full telephone number (as long as this is the caller's number and not a switchboard). For mobile calls, the location can be derived from GPS information. Trials in the US have shown that it is possible for the GPS information to be processed into latitude-longitude co-ordinates. (A cheaper system in which all raw GPS data is communicated proved too unreliable due to transmission problems). The related standard (WGS84) has been defined, but some further agreement is still needed, such as how many digits are required to give sufficient accuracy, and also how direction of travel should be coded.

Automatically generated calls follow a similar pattern. The modern systems used may also include information on the severity of the impact. Note also that such calls open a voice line 'just in case'.

4.4.3.2 *Emergency call prioritisation*

For GSM in Europe, provisions are already made to ascertain that any cellular phone may use any cellular network for emergency calls. Free roaming emergency calls are toll free and no billing procedures are used. What still needs to be achieved is for emergency calls to be recognised and prioritised. Within the 112 infrastructure, however, priority is assured.

4.4.3.3 *Language support*

Voice communication was found to be essential in Mayday services. There are various ways in which this can be supported in different languages. Whatever solution is selected, the support by automatic Mayday systems should be available to any operator involved. This requirement may affect the choice of operations. The solution proposed by E-MERGE is to involve Service Providers in a conference call with the driver's home Service Provider.

4.4.3.4 *Communication Protocols*

In Europe there are at present two proprietary protocols in use for sending such information to the PSAP. One is British and uses the protocol MLP Lite 112/999 (Mobile Location Protocol Lite), the other is Spanish and uses the protocol POSIC112. Both are derived from the more general LIF MLP defined in 2002 by the Location Inter-operability Forum, which, while waiting for the results deriving from the work by ETSI, seem destined to become the European protocol for the exchange of location between telephone operators and the PSAP. Due to the lack of a definitive European protocol it has not been possible to integrate the testing of this functionality with that which was developed by E-MERGE and E-CALLS. All this means that the management of location data from the mobile network is an open element for future research and experimentation.

4.4.3.5 *Crash/Impact Data*

The in-vehicle sensors in New York's ACN tests were able to provide crash information of use for determining appropriate field response. This included crash type and severity (e.g. indications of rollover and principle direction of force, deceleration time-histories, and the final resting position of the vehicle). However, the test results showed that to be valuable for impact assessment, the sensor system must be far more complicated than a simple airbag trigger. This requires a vehicle-based solution and, therefore, has to be left to the automotive industry. For the foreseeable future, this information will be an issue only for systems run by Service Providers. The use of a webcam in the vehicle could also be useful, but is currently very costly. Another possibility is that offered by the MMS facility of mobile devices.

4.4.3.6 *Good Samaritan Calls*

Such calls can be problematic, as they create a severe workload problem for PSAPs. With the rapid penetration of mobile phones, the number of Good Samaritan calls is rapidly increasing, typically reaching 40 calls per incident. An initial solution suggested was to introduce a 'Good Samaritan Button'. Pressing this button starts a call to a PSAP that is coded as a Good Samaritan call and includes location information and caller ID. This leads to notification - e.g. an icon - appearing on the PSAP screen map. After examining the implications of this approach, E-MERGE has proposed an alternative solution which involves the implementation of the ability to distinguish Good Samaritan calls at PSAP level.

If the operator is already well informed about a given emergency, no manual action is needed and the call is automatically acknowledged. If the operator is interested in having further information (which certainly will be the case where this is the first call coming from a site), a voice link can be opened from the PSAP and contact with the caller established. If no automatic acknowledgement is received or no voice link opened, the Good Samaritan call will automatically be retried.

4.4.3.7 Mismatched boundaries

Another problem is the lack of one-to-one correspondence between the coverage of wireless communication networks and PSAPs. If cell-ID is used, the communications network provider uses a mechanism to locate the caller, and then uses this information to select the correct PSAP. If the communication network provider is unwilling to provide this service, the cell-ID information is sent to the nearest PSAP. The result is that an emergency call from a cell phone may be directed to the wrong PSAP. In the European context, this raises a further potential difficulty, since the PSAP could be located in a different country and lead to a language problem (as no translation support is currently provided). The alternative is for detailed location information to be used, but this will take some more time. Both PSAPs and private Service Providers need reliable databases to reference PSAP boundaries.

4.4.3.8 Standard Interface

One of the main responsibilities of the PSAP is to co-ordinate the intervention of the rescue services, but such operative procedures and the interface for communication of details of the emergency to such centres is still undefined. Having carried out a first analysis of the emergency at the level of the PSAP, the information must be sent to the rescue centre on the place and type of accident, in other words, where and what kind of event is involved.

The definition of a standard interface at national level for the format and mode of exchange of information between the PSAP of first and second level would be of undoubted benefit to the efficiency of the emergency management chain and also in terms of co-ordination of the rescue services, especially for multiple cases in which several different types of rescue service are needed, e.g. fire service, ambulance and police. This last aspect, linked to the co-ordination of the use of modern technologies for carrying out these functions, is an open field for future developments regarding safety.

4.4.3.9 Handling and exchanging location reference information

A serious deficiency that emerged during Mayday and similar tests is the mismatch between location information coming from a GPS receiver and the street address given by a digital map on the basis of that GPS information. Mismatch between GPS location information and that of maps of the WGS standard is not a problem, since programmes for correction are generally available, but in roughly 1 out of 10 cases the address information proved to be seriously wrong. This means it is necessary to correct the maps used to handle Mayday calls.

The digital maps being used today are in most cases based on maps that were not made to give an accurate latitude-longitude position. For vehicle navigation, the map matching process allows for deficiencies that can be corrected with the help of information about the trajectory being travelled. For most other applications it is relative location that is important, not latitude-longitude location. Most applications can live with these deficiencies, but they make a reliable Mayday service impossible. The issue of map matching is discussed further in section 7.2 on radio navigation.

4.4.4 Future opportunities

Over the past ten years, considerable progress has been made towards establishing a common procedure for automated emergency calls. The basic legislation is in place, many of the technology issues have been resolved, the essential information requirements have been identified, and the principle features of the necessary architecture defined.

Among the remaining technical problems involved are the coding of the location of an incident, the provision of information on the nature of the impact in the case of crashes, and the handling of calls (especially the routeing of calls) in foreign languages. There appear to be good prospects that these will be resolved by current EC initiatives (e.g. via the e-Safety Forum or RESCUE project).

The biggest challenge for the future, however, is the actual implementation of these systems throughout Europe. Success will depend to some extent on action at European level, but most of all requires national commitment to making the necessary organisational arrangements.

Most of the developments involving provision of appropriate information regarding vehicle-related emergencies are dependent on the co-operation of the automobile industry in making it possible to generate the necessary linkages between onboard sensing devices, location referencing and communications systems.

ROSETTA, in line with other EC projects such as E-MERGE, therefore recommends actions to favour the co-ordinated adoption of the proposed emergency call architecture. This requires the following:

(a) Actions by all Member States to ensure the full implementation of e-112;

(b) Upgrading of the e-112 solution by PSAPs in order to enable them to handle the minimum set of data identified by E-MERGE;

(c) Measures to create necessary commitment from both private and public stakeholders.

4.5 Enforcement in ITS

4.5.1 Vision

In order to maintain safety in a road network it has become increasingly necessary to encourage and enforce certain standards of driver behaviour. Reduced speed and more moderate behaviour will also produce a much wider range of social benefits. For enforcement to be accepted, it must be seen to be both needed and uniformly effective. The vision is of a system of enforcement of speed limits and driver actions resulting in driver behaviour which enables policy targets of safety, capacity and reliability to be achieved. On-vehicle and infrastructure based ITS systems are at the heart of new more subtle and targeted enforcement approaches.

4.5.2 *State-of-the-art*

Enforcement may relate to moving or stationary vehicle offences and be introduced to improve traffic safety, operational efficiency, support priorities, improve the environment or recover costs. Particular application areas of enforcement using ITS include urban traffic management and parking, speed, lane change and access, as well as the payment of fees for public transport or for road user charging.

As telematics has replaced enforcement by police or traffic wardens, the enforcement has become more effective and the scope for individual tolerance has reduced. This has aggravated the ill feelings which often accompany being caught. Thus enforcement is a social and political problem, and one which not only requires technical solutions, but also political will and fair and practical implementation of institutional procedures. Enforcement is only necessary to reinforce education and engineering approaches which should have ensured that road users are aware of the need for enforcement in any particular situation. There is an additional concern relating to the high proportion of foreign traffic on major European roads and within cities, as enforcement must be applicable to foreign as well as national drivers.

Various automatic enforcement technologies are available. Many have been developed and tested within the European Framework Programmes and co-operation between the different groups is in place at many levels.

Cameras using wet film processes have been in use for several years to capture the number plates of drivers ignoring the red light at traffic signals i.e. 'red light running'. In some countries a picture of the rear of the vehicle with its licence plate is sufficient evidence. For this, a relatively simple arrangement with a pole-mounted camera upstream of the signal may be used. Other countries require that a picture be made that shows the face of the driver. This requirement complicates matters but it does not make automated capture impossible.

Issues of privacy and personal integrity have been particular difficulties in implementing particular camera based technologies, and still are in many countries. However, automated capturing of offences and automated processing have proved to be very effective in combating unsafe driving and reducing accidents. This has led to a greater awareness that legal arguments against automated enforcement should not stop the use of license plates as a means of tracing offenders, either directly or indirectly.

Other early applications of license plate recognition systems were the monitoring of vehicles at toll booths to check against a list of reported vehicles. Currently, for speed enforcement, most camera-based systems measure speeds at a particular location and the number plates of offending vehicles are captured. Traditionally, wet film processes were used which required sites to be visited regularly to change the film. The constraints on collecting film and managing cameras meant that at any time many camera boxes were unavailable to capture offenders. This has been largely overcome by the use of digital cameras and sophisticated analysis software. Whilst some cameras, such as those used for red light running offences, are fixed by the specific location of the offence, speeding may also be tackled by measuring elapsed time over a significant distance. This is perceived as much fairer than speed traps at specific points. It also affects traffic behaviour in a positive way by smoothing the flow.

In the Netherlands, speed enforcement using radar traps on motorways with a normal maximum speed of 120 km/h has reduced the number of violations from 20-40 % to 6 %. The use of trajectory speed enforcement brought this down further to less than 1 %. It should also be noted that those that were caught were mainly heavy goods vehicles for which the maximum speed is 80 km/h. Speeding by such vehicles could not be enforced with radar traps whereas the trajectory enforcement can separately identify such vehicles. As a result of these enforcement measures the number of accidents involving serious injuries was halved.

Other examples of automated video enforcement are the monitoring of the use of dedicated lanes for buses and/or heavy good vehicles or the monitoring of the transport of heavy goods. In combination with weigh-in-motion (WIM) systems automated video can be used to enforce maximum load and axle weight rules.

There are also projects on automated policing of urban parking zones and urban congestion charging zones using Closed Circuit Television (CCTV) systems. The cameras read the number plate which is then checked automatically against a database of vehicles for which payment has been made. Fines are subsequently issued to those not complying. The London Congestion Charging system is a recent example of such a system.

Modern automated video technologies for licence plate recognition can be efficient and cost-effective. As processing is automated, fines are received shortly after the violation. This means that violators can recall the situation, which improves the impact of the enforcement.

In-vehicle systems and co-operative systems with the infrastructure are also under development with the objective of automated intelligent speed adaptation. Such ISA systems may make enforcement much more acceptable, since it is technically possible to help the driver to avoid speeding. This can be done by warning, by tactile information (the throttle pushes back) or by fully automated control. Today there is no single fully operational solution to inform the ISA system in the vehicle about the prevailing maximum speed, but these are being developed.

Increasingly, police use roadside and vehicle-based video number plate reading systems to scan for vehicles used by known offenders and to check that vehicles are properly licensed and insured. This requires that the number plate is accurate and belongs to the vehicle. Automatic Vehicle Identification (AVI) systems are being developed to overcome the growing stolen and cloned number plate problems and may form a future basis for enforcement more generally.

4.5.3 *Issues*

It can be observed that enforcement systems for road user charging can work very efficiently due to the financial implications for the operators and the corresponding high penalty charges. There will be many more applications that will be possible for traffic control.

However, a different situation concerns automated speed enforcement, because of the ubiquitous nature of the problem and because control philosophy is still not universally accepted socially and politically. In many countries automation is still in its infancy, mainly because lobby groups have stalled the necessary changes in legislation. However, the effects on safety justify an active approach to overcome the social and institutional barriers that prevent the application of modern (e.g. license plate based) enforcement. The successful development of efficient speed enforcement also has a positive effect on traffic quality and traveller comfort, and additionally has received public support where clear evidence of an improvement in safety is available.

Opportunities for the future are in the active promotion of new enforcement techniques accompanied by institutional agreements and support actions, which include:

- Building on from the Memorandum of Understanding (MoU) recently agreed by the European police forces for equipment approval based on the Schengen Treaty applicable for the full-scale cross border enforcement domain. This successful result has led to co-operation across European borders between traffic police forces under the name TISPOL, such that violators caught in a foreign country will be prosecuted in their home country;
- For the immediate and near future, action on harmonisation of legislation should be undertaken by the EC;
- A mechanism for unified cross-border communication and information exchange needs to be considered in order to cope with mass communication and ensure the effectiveness of processing and compliance with cross border enforcement requests;

- Type testing and type approval is required to ensure that digital enforcement systems are accepted by legal institutions and ultimately, by drivers. It is of primary importance for cross-border enforcement that certification of equipment is granted at European level, and not only at national levels, as now;
- Ways to safeguard the privacy of individuals whilst maintaining efficient and effective enforcement need to be developed;
- Enforcement policies need to be reviewed in conjunction with the advancement of technology to provide much greater opportunity and subtlety in application.

Much of what is concluded here is in line with recommendations from the VERA project as sent to the EC and European Parliament. ROSETTA commends an active promotion of new enforcement techniques and the necessary support actions.

4.5.4 *Future opportunities*

The value of modern automated enforcement is well understood by the various police forces working together in the TISPOL arrangement. However more support is needed and for this research must provide additional data on the impacts.

The various vehicle and vehicle-infrastructure systems which are being developed and deployed contain location and communication technologies which could be used for enforcement. Following an accident, the police have the right of access to all the evidence and this may include speed information from the engine management system and from a navigation system. Future applications will bring huge opportunities to enforce and control, and a balance between privacy, enforcement and security must be struck.

An area of future enforcement will relate directly to security. Already in the U.S. some trucks are fitted with electronic devices which automatically disable the vehicle if it approaches certain sensitive locations.

Chapter 5

Network Management

5.1 Traffic Management and Control

Today's traffic systems are highly complex. A great variety of public and individual transport modes like buses, light rail, rapid transit taxis, trucks and cars, as well as cyclists and pedestrians, often share the same limited space of the transport infrastructure. Every user has different trip purposes, speeds and destinations, creating a great mixture of origin and destination movements, often interacting with each other. On top of this, traffic demand is still increasing, while the possibilities for new road infrastructure in particular are limited due to environmental, social and financial constraints. A key question is how to cope with these challenges without building more and more road infrastructure?

A significant answer is traffic network management and control. Together with integrated re-design of existing roads or the building of new road infrastructure, it can be used to manage cars and goods transport for a more efficient and responsible use of road space.

In recent decades the common problem of traffic congestion accompanied by high economic, environmental, and social losses led to a concerted development of a variety of promising solutions all over the world. However, many of those approaches were only implemented on local test sites and remain largely unknown elsewhere. The reasons for this are the inadequate dissemination of results, the lack of best practice guidelines, the lack of public authority funding or the tendency to continue with known fail-safe conventional systems. Another fact is that every region, area or city presents a unique context with its specific structures, public transport modes, systems and traveller behaviour.

This calls for new initiatives and opportunities to benefit from relevant research through more effective propagation of results, and their applicability with respect to the prevailing environment, demonstrated through best practice guidelines as well as cost-benefit analyses.

5.1.1 Vision

A rather extreme vision of future traffic management and control is one where, particularly for densely populated areas, the traffic on the main arteries would be almost exclusively operated/controlled by fully automated public and private transport modes. Without human interference, systems will ensure that the automated transport is predictable and perfectly harmonised for passengers and freight alike. This kind of vision, however, would assume a complete change of today's transport infrastructure and also serious restrictions on the right of free movement, especially for passengers.

A more realistic vision and the general objective of any traffic control or management system is the well adjusted integration of all systems to permit a highly efficient, safe and harmonised operation of all modes within and between the urban and interurban networks.

A basic prerequisite for this integration is to obtain a fully detailed and all-embracing overview of the general status of every mode in the network, accompanied by supplementary information on incidents, road works, emissions and weather conditions. This can be realised, for example, through a data-warehouse concept collecting data at all detection, planning and operational levels. The traffic information/management centre processes, analyses, develops, visualises and finally disseminates the current network status information for strategic planning, as well as for short and medium term prognosis.

Within this vision public transport is widely promoted. Access to and between public transport and all other transport systems is facilitated through well designed transfer points. Public transport supply is adapted to traffic demand.

Measures are deployed to manage efficient freight traffic on major arterials as well as to control loading and unloading of goods in urban areas.

Information on the current status of the urban network, traffic conditions, travel time on major links, availability of nearby parking space and alternative public modes is provided through collective media (e.g. variable information boards), but also through individual means of information such as personal traveller assistants or systems within vehicles. This information is consistent with the choices of the traffic management, so that an overall efficient equilibrium is maintained.

When there is limited road space or variable transport demand, more advanced telematics systems are deployed. These include flexible operation/management of road space and lanes through guidance provided on overhead panels, flexible lane markings integrated in the road surface or advanced in-vehicle systems, visualising lane marking or currently accessible lanes for the driver.

Urban areas suffering from severe environmental impacts use special sensors for local monitoring of exhaust fumes. This data enables on the one hand real-time and preventive adjustment of actual strategies for general urban traffic control, and on the other hand the detection of individual environmentally critical vehicles, which may then be denied access to sensitive areas such as city centres.

For maximum flexibility all components of the traffic management system are based on an open system architecture with standardised interfaces. Roles and responsibilities of stakeholders, planners and operators are clearly defined, and perfectly trained staff are available and familiar with the system and its impacts.

An essential prerequisite for all advanced systems is the availability of safety systems in case of malfunctions or operation under degraded conditions. The performance of the systems is constantly assessed and validated by Total Quality Management (TQM).

5.1.2 *State-of-the-art*

Numerous solutions for specific issues concerning traffic control have been developed and successfully deployed in urban and interurban areas and have international acknowledgement. A selection is given here.

5.1.2.1 *Traffic monitoring*

Important progress has been made in the area of traffic monitoring, a prerequisite of any traffic control. A more detailed look at this is given in section 0. Urban and suburban networks are increasingly equipped with detection loops, overhead sensors or other advanced detection measures such as Closed Circuit Television (CCTV). In addition Floating Car Data (FCD) from individual or public vehicles are used to measure the real-time performance of the network. Also information on weather, road works, environmental issues or simply on periods of increased transport demand, such as at holidays or public events, provides valuable assistance. An example of advanced environmental monitoring is the 5th Framework Programme project HEAVEN. It produced real-time air quality monitoring and modelling applications feeding into a decision support system. Furthermore, it established a data platform for the assessment of emissions and health effects for air pollutants and noise caused by traffic.

5.1.2.2 *Control measures*

Contemporary urban control measures operate both across the whole network and at a local level. Measures vary from a simple display of information to travellers to explicit control measures, such as prohibiting access to an area by motorised vehicles during smog weather conditions.

5.1.2.3 *Soft measures*

High-quality overall network performance is achieved through the balanced distribution of traffic demand to the available supply. Organisational approaches focus on influencing mobility behaviour using soft measures, such as campaigns giving mobility advice to individuals, pupils, specific traveller groups or companies, e.g. in the European project TAPESTRY. This 5th FP project aimed to encourage European travellers and goods operators to adopt a more sustainable, intermodal travel behaviour by providing recommendations, guidance and practical advice on the potential of multimodal travel awareness campaigns and their cost effectiveness.

5.1.2.4 *Sustainable measures*

Here also the successful Eureka programme of PROMETHEUS and DRIVE/DRIVE II or the CIVITAS initiatives have fostered sustainable international and interdisciplinary work combining traffic engineering, vehicle engineering, economic and social science. Though not exclusively dedicated to urban traffic control, CIVITAS addresses ambitious cities that are willing to introduce sustainable urban transport policy strategies. Approaches include new demand management strategies, innovative logistics services or integration of transport management systems. The DRIVE programme covered projects dealing with demand management, information systems, integral management of urban traffic and public transport management.

5.1.2.5 *Adaptive control measures*

Traffic-adaptive algorithms and adjusted linkage of signal controlled intersections are still topics of research in order to enhance the performance on a route or across a network. Today's advanced systems focus on the use of current detector data to modify control online. Several systems have been developed in Europe and elsewhere with different control philosophies and degrees of local and central intelligence. Adaptive systems use model-based dynamic estimation of traffic flows enabling optimised traffic control. Examples of adaptive systems are SCATS/SCOOT, SPOT/UTOPIA and BALANCE. The project SMART NETS aims to improve the state-of-the-art real-time network-wide urban traffic control via the new-generation control strategy TUC (Traffic-responsive Urban Control). Improvements of up to 40 % in journey times as compared to fixed-time settings under saturated traffic conditions are expected. The concept is being tested in several European cities. Another example is the project SURF 2000 using both macro- and micro-regulation to optimise the overall network, and which includes local actions on intersections in a zone of approximately 40 intersections. Several strategies have been defined for vehicles and pedestrians.

5.1.2.6 Public transport priority

Priority for public transport is also part of the advanced management of intersections and often serves as a path to adaptive network control. The 5th FP project PRISCILLA focused on the dissemination of best practice guidelines for bus priority in wide areas (100-300 buses, 80-130 intersections, city centres and suburbs). PROMPT developed and demonstrated techniques for giving active priority to buses and trams in fixed and real-time adaptive urban traffic control systems. Many other online network applications for bus priority have been piloted in projects of the Telematics Applications Programme.

5.1.2.7 Access control

Access control systems or city pricing are deployed as effective measures to reduce entry into specific areas of a city, e.g. in Barcelona, Bologna, London and Singapore. The revenues enable better public transport services to be offered as an alternative to individual transport.

5.1.2.8 Goods transport

Increasing goods transport in urban areas initiated the development of measures like freight management centres and strategies for freight/fleet management or for sharing transport capacities (see also Chapter 6). Many projects derived from the CIVITAS initiative developed new approaches for innovative logistics services, vehicle routeing or loading/unloading area reservations. The 5th FP project MOSCA aimed at the development of a tool for integrated planning and control of production and transportation processes (such as booking and reservation procedures, vehicle routeing, loading/unloading area reservations, emergency management support, efficient multimodal inter-connection) in a single platform.

5.1.2.9 Parking management

Problems of freight transport and delivery in urban areas are often directly linked to parking. Repeatedly on-street parking of vehicles can disturb loading and unloading of goods and vice versa. Parking guidance systems, as deployed in the DRIVE II project LLAMD, in many national projects or in urban parking management, minimise the search for parking space and prevent uncontrolled parking which often negatively influences through traffic. Restrictive strategies proved to be quite successful, also fostering an increased use of alternative modes, e.g. public transport. The 5th FP project E–PARKING introduced new user-friendly and safe technologies in the booking and selling process of parking spaces in order to increase competition and productivity.

5.1.2.10 Information services

Soft means of network management include collective or individual information services. Information on network performance is made available through radio traffic information via FM voice radio or RDS/TMC services, Personal Traveller Assistance (PTA) or in-vehicle navigation systems.

5.1.2.11 Integration of systems

Strategies were also developed for the integration of urban traffic control (UTC), Driver Information Systems (DIS) and Public Transport Systems (PTS) within an urban traffic management system e.g. in the 4th FP projects INCOME or MUSIC. Within these projects novel control methods were evaluated for network traffic management, alone or in combination with other measures such as park and ride, re-allocation of road space to public transport, road pricing, etc.

5.1.2.12 Variable Message Signs

For network control, Variable Message Signs (VMS) give advice for best routes within or through the interurban network, but also in regional and urban networks. An example is the German project MOBINET. Different types of graphical displays are used to direct traffic from the motorway network and to advise on the quickest route to the city centre. The systems 'Netz-Info' and 'Ring-Info' display the current traffic status on major arterials in Munich without proposing a mandatory route. The signs are especially developed for daily commuters familiar with the road network and thus enable the drivers to individually choose their best route. In Greece, the APOLLO system measures roadside air quality and predicts the progression of pollution. The city's traffic managers can divert traffic away from the central congested areas before a serious crisis strikes. The cities of Barcelona, Berlin or Hamburg also use variable allocation of lanes for flexibly adapting the number of lanes per direction to the current traffic demand.

Motorway network capacities are achieved through the distribution of traffic flow over space or time predominantly via variable message signs for link and network control. Travel times to exits are successfully shown using VMS on the Boulevard Peripherique in Paris and also for different alternative routes on Dutch motorways. Link control increases traffic safety and leads to harmonisation and increased efficiency of traffic flow, predominantly through speed control. Information on road works is a major field of application of network control. DRIVE II projects have demonstrated the benefits of network control. The management of regional networks was tackled in the past by LLAMD and, mainly for the metropolitan case, by QUARTET which focused on the implementation of a fully integrated road transport environment using Advanced Transport Telematics (ATT) technologies.

5.1.2.13 *Ramp metering and incident management*

In RHYTHM of the 5th FP the use of video processing was tested for combined ramp metering and road surveillance. Ramp metering for secondary road network entries (EUROCOR) as well as variable allocation of lanes and the temporary use of the hard shoulder have been deployed via overhead VMS. The reservation of exclusive lanes for buses or taxis, especially during peak hour periods, fosters greater use of public modes.

IN-RESPONSE and then PRIME in the 5th FP devoted attention to incident management, prediction of accidents, etc. In PRIME, congestion and accident prediction were analysed to inform traffic management.

Several studies were aimed at improving the algorithms for flow control, ramp metering, and network control via re-routeing. Simulations and practical experimentation have demonstrated their benefits. Prototypes and medium scale network control systems were then deployed as part of the Euro-regional projects in the TEMPO programme.

The GERDIEN project promoted an overall architecture for motorway networks, encompassing the various telematics measures.

5.1.2.14 *Interfaces and architecture*

Important support actions for network management are standardised interfaces and clearly defined data structures, as were defined by system architecture projects like KAREN and FRAME, as well as the CEN TC 278 working groups.

5.1.3 Issues

Network management and control concerns both urban and interurban environments. Urban street networks are embedded within the regional environments of the highway and motorway networks. The urban and rural road networks also provide supply to public transport systems. In addition, they are linked to public transport via park and ride, passenger stations, and traveller information systems.

Interurban areas include at least two types of networks. The first type has low capacity and aims to grant accessibility to sparsely populated zones. The second type is the high capacity motorway network interconnecting metropolitan areas. So far, for sparsely populated zones, there is almost no alternative to the car, whilst for the motorway networks the predominant alternatives are railway and air transportation. These network interdependencies require comprehensive transport planning and a co-ordinated traffic management of the infrastructures. Attention has also to be given to the interface between urban and interurban areas or between bordering interurban areas – areas that have to adjust to or in the worst case buffer the bordering technical, operational and institutional demands. The priority issues for a greater penetration of ITS measures in urban and non-urban networks are considered below.

5.1.3.1 Strategies for integrated urban traffic management

Integrated traffic management is required to enable cities and regions to cope with the tasks and challenges of future transport. This requires co-ordinated traffic control strategies at all levels of transport planning and operations. In Figure 8 an attempt is made to show how traffic management and control should be considered as part of urban and regional transport infrastructure planning when scenarios of future sustainable transport development are being studied.

Strategic traffic management plans are to be developed at an inter- and multimodal level. Strategies for dynamic control of the road infrastructure are to be developed to meet the tasks of recurrent and non-recurrent events in traffic. Traveller information systems are either public tasks or provided by private service providers.

A major effort in this context is the integration, linking and extension of existing systems and novel urban control systems, including the communication platform between and within the individual systems. This technical integration is a prerequisite for enabling the telematics infrastructure to serve the individual telematics measures.

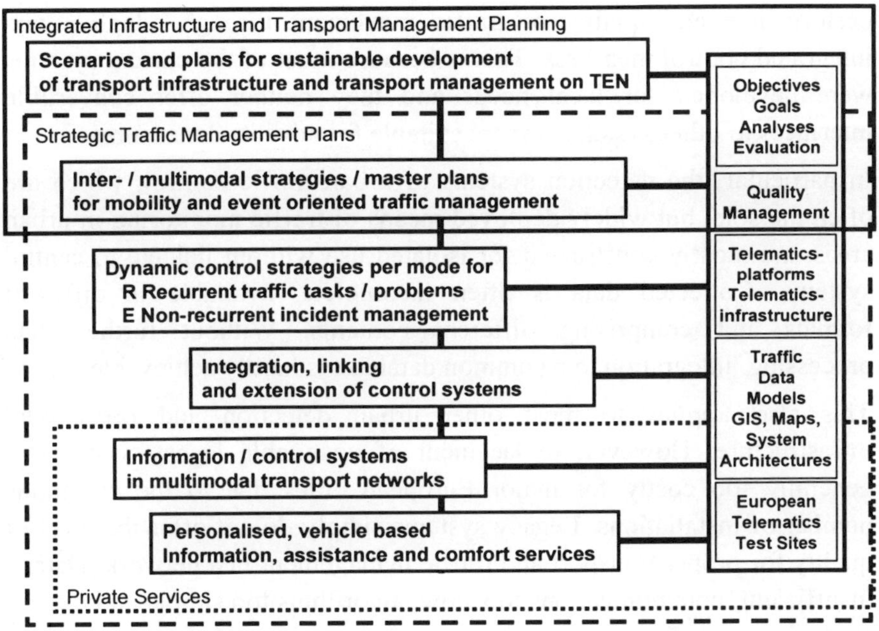

Figure 8: Traffic Managements Plans within the Hierarchical Structure of Transport Infrastructure and Transport Planning (Keller, 2004)

But the strategic and technical requirements of an integrated urban traffic management are also to be developed in consideration of further aspects. These are e.g. the objectives of the stakeholders, the evaluation of the consequences of associated measures as well as a total quality management of the elements in the information and control value chain to ensure adequate and constant quality.

5.1.3.2 Enabling urban detection and communication systems

A major challenge for the deployment of novel and integrated traffic management and control systems is to enable the existing detection and communication systems of the cities to be able to co-operate with and serve the new telematics systems.

This applies in particular to the urban traffic signal systems. Currently, conventional systems dominate, hampering the achievement of better performance and quality of traffic movement through adaptive and integrated control measures. This is because most of the current systems were developed for local needs and they neither offer convertible interfaces to other systems nor are suitable for advanced measures.

In particular, the detection systems of urban traffic control, which are often the sole, but widely deployed means of traffic monitoring in urban areas, are mostly constructed for isolated use without linkage to central systems. Collected data is often incomplete, available in different formats and comprising different contents. Without further data processing, integration to a common database is hardly achievable.

This also applies to most other urban detection and operational infrastructure. However, replacement of unsuitable legacy systems is generally too costly for major European cities due to the enormous number of installations. Legacy systems simply do not offer the level of quality for proper transport and traffic management in a network. Data is insufficient, communications too slow, algorithms too simple and overall strategies non-existent.

Other challenges also exist for advanced systems. In the case of malfunction, operation in fallback mode often can lead to extremely degraded performance compared to conventional systems. In particular, malfunctions of some advanced approaches, like substitutive variable message signs or flexible lane allocation, can result in severe orientation or safety problems.

5.1.3.3 *Management of regional, large scale networks*

Research on network management has been important for the solutions to specific problems, e.g. flow control, ramp metering and dynamic routeing in networks, where the basic elements are known sufficiently. In parallel, proposals for overall architectures have been made and are now part of the European ITS Framework Architecture. However, the problem of consistent management of large, regional networks in different traffic conditions, encompassing a set of measures and considering the different jurisdictions has not largely found accepted solutions.

It can be considered in principle as an architectural problem by defining a suitable architecture, with attention to the organisational part. As such, it will be relevant for many countries, but a convenient solution should be sought looking at the emerging countries. It is also a problem of traffic management strategies. New, suitable strategies have to be researched.

With the increasing importance of mobility and with the high level of demand for road traffic, the ability to manage network-wide events with the required level of safety, efficiency and comfort for the user becomes more and more important. Predictable and relatively frequent events are normally dealt with at some level by existing management systems, e.g. snow in Nordic countries or in Alpine regions. Less frequent and less predictable events have had less attention, but these can also result in large blocks of traffic in entire regions and/or in unsafe conditions.

5.1.3.4 *Cost of ITS deployment*

One of the reasons for the very limited penetration of the ITS applications in rural networks, but also in urban and regional networks, is the cost of such applications. Costs are especially high for:

- Monitoring which requires installing many detectors;
- Communication which requires extensive networks;
- Information to users, which is normally done via VMS, again expensive and cumbersome to install;

- Operations, especially maintenance of detectors and information systems.

For a larger penetration of ITS, research and development effort should aim at reducing those costs.

Cost reduction is unlikely to come from market expansion. For the electronic components, the market of non-urban traffic management will always be too small. Moreover, a dominant cost component is represented by the installation cost, which will not be touched by any possible market expansion.

The following strategic lines should be addressed:

- *Using personal or in-car devices to distribute information*

  One of the larger cost components is the Variable Message Sign – expensive in itself and expensive in installation. With the increasing penetration of personal and in-car devices, the need for important VMS installations can be reduced. In turn, this requires research on the possible use of personal devices and on the efficient coupling of on-street devices with personal and on-board devices, looking at the overall system robustness.

- *Using emerging technologies for detection, with attention to wireless communications and sensor networks*

  Wireless technologies are developing impressively and could contribute to reducing cost and increasing efficiency of complex architectures. GERDIEN (motorway) and QUARTET-like (urban) architectures, where intelligent outstations connect various detectors and actuators in a limited spatial environment, can be boosted by a tuned use of wireless connections. In another context, rural systems can gain from low-cost, generic wireless communications.

Personal information services are expanding even if not at the rate expected by optimistic forecasts. They can also benefit from traffic monitoring. One way to reduce the costs of traffic management is to share the cost, where possible, with personal services. It could be in the form of using the same basic traffic data for both purposes, complementing the different data sources or using data at reduced costs, e.g. using floating car data obtained from a personal service for increasing safety in rural areas.

5.1.4 Future opportunities

5.1.4.1 Integrated urban network management strategies

The major objective of urban network management is the development of an overall integrated intermodal control system. This requires joint elaboration of general strategies and measures by all concerned stakeholders. In this context, future opportunities include:

- An ITS policy framework architecture with clear allocation of competence, responsibilities and financial involvement for development, deployment and operation of network management to minimise deployment problems;
- Instead of costly fundamental research or development of new urban control approaches, efforts should be especially focused on detailed analysis of urban conditions and requirements to effectively combine measures for an overall integrated system;
- Based on these framework conditions, strategies for integrated network management need to be developed ensuring a sustainable transport operation, especially in view of the regulations for clean air;
- There is a need for self optimising, adaptive control systems flexibly reacting to varying traffic demand. The approach to flexibly control and integrate multimodal systems within one urban network still needs to be found and for that reason research activities in the field of simulation should be further pursued;
- Because advanced systems are mostly expensive, elaboration of economic feasibility, cost-benefit studies and examples for Public-Private Partnerships (PPP) need to be encouraged. The long-term goal to be achieved is the decrease of operational costs for improved urban network control through sufficient

operation and maintenance, as well as through fair dissemination of costs ('polluter pays' principle) as demanded in the 'White Paper – European transport policy for 2010' (European Commission, 2001).

Further research is necessary in the field of a common multimodal urban system architecture. Not specific for transport research but an elemental issue are open system architectures, standardised infrastructure and data formats as well as satisfactory bandwidth for data transfer.

5.1.4.2 *Quality management of urban data and communication*

For overall monitoring of network performance, collection and integration of all available online data is especially important. In this context the challenge for the deployment of novel and integrated traffic management and control systems is to enable the existing detection and communication systems and to link them with new ITS technologies. This requires:

- The development of universal adapters and upgrade kits for legacy systems, including concepts for isolated devices, to transfer data to central units and to reduce costs, i.e. for new detection infrastructure. Particularly in areas with a low density of detection infrastructure this must be supported by new tools and algorithms for the simulation of traffic demand;
- Adequate quality/performance control measures and algorithms to make possible the high-quality integration of various systems into a common shell, presupposing common standardised infrastructure and data formats;
- A total quality management ensuring constant validation of performance, e.g. via a quality index to increase user acceptance of systems. The quality index is also a valuable tool for the calibration and improvement of traffic control systems.

Finally, new approaches for urban traffic control will create new situations amongst traffic modes and surrounding environments, as well as for road users. Surveys on possible interactions have to be initiated to minimise negative impacts like disturbed traffic performance, degrading of environmental conditions or uncertainties of road users when approaching or moving within the system.

5.1.4.3 *Regional, interurban co-operative traffic management schemes*

With the increase in demand and the limited opportunities for increase in supply, the management of the complete traffic network, for efficiency and safety, becomes crucial. Management must be capable of integrating urban networks, the motorway network and sparse rural networks in a consistent set.

In this context, management is, by definition, co-operative. All networks have to co-operate, because congestion expands through network boundaries – and accidents migrate as well. But co-operation is also needed between infrastructure and vehicles. Without this co-operation, efficient strategies will be too difficult and/or too expensive.

This fact is particularly true when considering non-urban networks. For those networks the following recommendations are applicable:

- Expand the research and the demonstrations into regional, co-operative traffic management schemes, looking both at the architecture and continuing the work of the European Framework Architecture, as well as at the functional issue, i.e. models and algorithms for managing a network of networks;
- Extend the research on management of networks under abnormal conditions – where such conditions could be due to excess demand, major changes in demand or to relevant modifications to the supply, e.g. for weather conditions or large accidents;
- Research on efficient integration of personal information services and management both for reducing cost in deployment and operations and to increase effectiveness;
- Research on the possible applications of new technologies to traffic, e.g. combination of wireless and sensor networks.

One of the possibly relevant outcomes of the research in nano-technologies is the development of the sensor networks ideas such as smart dust. Clear advantages could come from the combination of the two technological developments. The local sensor networks could be combined with wireless networks and finally with wireline backbones to offer a complete architecture for motorway or rural traffic management, obtaining benefits from the deployment of personal information services.

5.2 Road User Charging

5.2.1 Vision

Intelligent Transport Systems (ITS) have the potential to transform Road User Charging (RUC) into a powerful instrument to support European transport policy objectives. However, the application of sophisticated charging structures, which reflect detailed characteristics of road use and can offer effective tools of traffic management, require strategic decisions at European level on interoperability of the technologies and procedures adopted. If road user charging were to become widespread, basic telematics would need to be installed in most vehicles, and this would generate additional opportunities for other market-driven ITS systems and services.

Road users have traditionally been charged on the basis of vehicle ownership through purchase and registration taxes and on usage, mainly in the form of fuel tax. Road user charging is a concept in which charges to the motorist are directly related to use of the road network and can be calculated on the basis of a range of factors, such as the distance driven, type of road, traffic conditions, time of day and vehicle type. Such charges may be used wholly or partially to replace current taxes and therefore, may be revenue neutral (in the sense that they address non-fiscal objectives) or may include an additional charge to manage demand in some way or to generate revenue for facilities or services. In the US, road user charging is referred to as 'Value Pricing', which reflects the idea that the road users are getting what they pay for. Unlike 'road user charging', the term implies a sense of commercial value or of benefit derived from the payment.

In the ROSETTA vision for road user charging, all vehicles will be subject to charges which can be varied by time and by location wherever they are in Europe. The charging regimes will be driven by clear policy objectives which will reflect increasing concerns on issues such as congestion, the environment, safety, economic competitiveness and social inclusion. Vehicle characteristics and journey purpose may also affect charges applied. In this vision, an architecture will exist which will enable a range of systems to interact with interoperable charging 'dialogues', which are sufficiently flexible to allow any road to be subject to any RUC policy. The following sections look at the current situation, the issues which need to be taken into account if ITS applications are to enable fulfilment of the vision and, finally, the necessary actions and research which the 6th FP might usefully address.

5.2.2 *State-of-the-art*

The EC policy statement on transport infrastructure charging (2003) states that 'one of the principal reasons for the imbalance in the transport system is that the transport modes do not in every case pay the costs for which they are responsible... the Commission's infrastructure charging policy is that transport taxes and charges, in every mode of transport, should be varied to reflect the cost of different pollution levels, travelling times and damage costs as well as infrastructure costs - to apply the 'polluter pays' principle and provide clear fiscal incentives to help achieve goals of reducing transport's congestion, pollution, re-balancing the modal split and decoupling transport growth from economic growth.'

The importance of sophisticated tolling technology to support policy objectives was seen in the 2004 European Parliamentary Directive 'On the interoperability of electronic road toll systems in the Community' (Directive 2004/52/EC) which applies to all electronic forms of tolling requiring on-board equipment. It states that:

- The work for deciding on a European Electronic Toll Service has now been launched (with a final decision to be taken by 1st July 2006);

- After the decision is taken, within three years the service shall be offered by all operators in all Member States for lorries and after five years for all vehicles;
- The permitted technologies include satellite positioning, DSRC at 5.8 GHz, GSM/GPRS and must be 'interoperable' at the European level;
- Electronic tolling should be promoted (with a target of 50 % of the flow by January 2007);
- Member States maintain full authority in deciding on policies for payment, since the Directive relates only to methods.

To date the most extensive systems currently in operation are those used in continental Europe for motorway tolling, and for collecting toll payments for particular road sections, such as tunnels. In these cases, the aim has normally been to generate revenue to finance infrastructure and its maintenance. Examples of such tolling schemes exist in France (TIS), Italy (TELEPASS), Spain (VIA-T), Portugal (VIA VERDE) and Austria (EUROPASS). Several schemes have been in operation for many years and payment systems have evolved from the use of paper (cash) payments, to encompass credit and smart cards, and electronic tolling. Payments are made at 'toll plazas' and are generally calculated at the exit on a distance basis for the network applications. In some cases tolls are used to collect a 'passage fee'. Current electronic tolling is based on DSRC with a mono-lane system where vehicles pass through toll gates, i.e. gantry systems, traditionally at toll plazas. More recently, the toll 'gates' are on the carriageways themselves as for free-flow tolling i.e. 'virtual' toll plazas consisting simply of one or two successive gantries. The systems used in different European countries involve different approaches and technologies.

Charging schemes have also been introduced on urban roads. These are most commonly a way of reducing congestion (e.g. Singapore), of generating revenue (e.g. in Trondheim, for major infrastructure investment) or both (London). Eight of the many urban schemes elsewhere in Europe were studied in the CONNECT project. They include electronic charging schemes using Dedicated Short Range Communications (DSRC), smartcard systems and paper based systems.

All are cordon operated and Automatic Number Plate Reading (ANPR) systems are largely used as a basis for enforcement.

Systems have also been developed specifically for heavy goods vehicles, largely to enable costs to be recovered nationally for international freight movements. These are dependent upon autonomous vehicle location (normally obtained via satellite positioning, possibly with local augmentation, and in some cases paralleled by inertial navigation or map matching), in-vehicle trip logging, and communication (GSM/GPRS) with a central service which calculates the fee and does the billing. These methods, which require on-board units, are supplemented by payment through self-service stations and/or the Internet. Examples are the German TOLLCOLLECT system designed for heavy goods vehicles in which the trip cost is calculated on the basis of log records sent from the on-board unit to the central billing service, and the Swiss LSVA in which DSRC is used to trigger at the entrance to and exit from Switzerland. A connection to the tachograph plus GPS is used to measure the distance travelled; log records are then transferred to the central service through a smart card.

In the USA, roadside beacon-based systems using dedicated shortwave roadside communication (DSRC) systems at 5.9 GHz are being developed as a standard for vehicle-to-roadside communication which would include applications of charging. The same frequency has been settled on for Europe for DSRC and some charging applications will adopt this technology where the enhanced certainty, quality and quantity of communication will be seen to outweigh the fixed point limitations and likely additional vehicle costs.

5.2.3 *Issues*

5.2.3.1 *Choice of technology*

Introduction and operation of a toll-based approach entails a small cost for the user but a high cost for the infrastructure owner. It is suitable for specific sections (tunnels, motorways, trunk roads), where toll plazas can intercept both entry and exit from the network. It is able to calculate the charge for the entire trip (based on time, length or other cost functions), so is suitable for large, but closed networks such as national motorway networks. It can also be useful for pricing options such as 'cordon pricing' (as in Trondheim) or Access Control schemes. As far as the user is concerned, the convenience increases with the dimension of the network, as payment is made just once for an entire trip. However, a large network is likely to require administrative agreements between many road operators (the largest in operation, in Italy, involves 26 motorway concessionaires) as well as a clearing house to apportion the fees paid. Also, as network size increases, the ability to selectively vary charge by location decreases.

The trip-based approach requires an on-board unit and can be used for more flexible road user charging schemes in which charges may vary by time and location. More subtle approaches to charging may reflect the costs of a journey for society – an application of the 'polluter pays' principle – and/or the benefits for the road user (Glaister, 2005). It can also apply any pricing policy operated by any toll plaza technology by using the concept of virtual toll plazas.

There are three technology issues associated with a trip-based approach:

(a) The lack of precision and robustness of satellite positioning in some circumstances. These shortcomings are most evident in urban areas;

(b) Difficulties in setting up and maintaining a complex distributed architecture with the necessary level of performance;

(c) A complicated system for the administration of post-payment is needed.

It is possible that with an improvement in performance of the technologies for automatic number plate recognition or the introduction of electronic vehicle identification, an architecture could be developed in which the vehicle is a completely passive element i.e. no on-board unit (OBU). In such a situation, the points for fee payment would no longer be 'toll plazas' but 'enforcement plazas'. An application of this architecture is found in the London congestion charging scheme, where vehicles are not required to install any equipment, payment can be made in different ways (including post-payment), a 'cordon pricing' policy is applied and vehicles are detected and number plates read (for enforcement purposes only) at the entrance of a closed network. However, it should be noted that, since vehicles are not required to install any equipment, then the London case will not come within the scope of the Directive 2004/52/CE. In general, this type of scheme should be considered as an intermediate stage to enable policy-driven short-term road user charging applications to be developed. Vehicle location and communication technologies and systems are developing strongly for navigation, safety/security and other purposes, such as usage-related vehicle insurance. When these technologies have further developed and matured they will be likely to form the basis for sophisticated trip-based charging systems. Issues which would need to be addressed in transition relate to:

- Transparency of and confidence in charging delivery mechanisms – e.g. GPS with Galileo will shortly offer significant increases in performance and integrity over GPS stand-alone;
- Different technologies have different costs and different cost sharing between the various sectors (e.g. between operators and users) which imply different contributions to other policy goals and potentially different levels of user acceptance;
- The need for a clear link between policy aspiration and price structure – for example, congestion is highly localised in time and space, so blanket pricing which calls itself 'congestion charging' would be met with scepticism, even resentment;
- The extent to which wider policy considerations affect user acceptance – e.g. social impacts, climate change, traffic and community impacts. The converse is the degree to which 'selling' road user charging as making towns and cities better places to live and reducing climate change will be labelled 'nanny state' intervention or simply mistrusted;
- The response in provision of alternatives to the car – walking, cycling, rail and bus;

- The impact of mode shift in response to road user charging on the acceptability of other modes;
- The extent to which media play on the issues will influence public acceptance and understanding.

5.2.3.2 *Enforcement*

For 'point-based' methods of charging, technologies for enforcement are well established using cameras and automatic number plate recognition.

For trip-based methods, where there are no fixed tolling points and the only information available is the log of the trip (TOLLCOLLECT and LSVA), enforcement is more complicated. Complete enforcement needs on-the-road verification of the on-board-unit (OBU) and, for non-equipped vehicles, verification of proper payment. For ITS, this implies that the OBUs must be capable of communicating with external devices. This may be automatic at set locations which detect and classify a vehicle and verify proper payments automatically, or an equivalent portable device may be available to the police or others. In examples in Switzerland and Germany, the vehicles are equipped with short range communication systems. Such a system may need to be capable of being used with various technologies and services.

5.2.3.3 *'Value pricing'*

Linnett (2004) envisages a wholesale shift in the way that charging technologies are used; highway authorities will no longer be mere providers of road space, they will be 'selling journeys'. In this vision, the opportunity exists to manage the sale of journeys in much the same way that airlines sell flights or that train operators sell tickets - with a mix of advance sales, pre-booking and, at a higher cost, turn-up-and-go. However, the application of such an approach to road travel would require a level of clarity of understanding of capacity and its management which currently does not exist. The integration of charging systems with information systems would not only defray the cost of the charging technology by adding commercial uses, but would also improve management of the infrastructure, encouraging people to travel when the infrastructure is lightly used in order to reduce charges.

5.2.3.4 *Privacy*

The better performing systems, which base the calculation and billing on the log record of vehicle movement must address privacy issues. It may possible for a pre-paid smartcard to be automatically debited, keeping a local (probably short-term and rewriteable) record of the journey undertaken to enable the owner of the vehicle to challenge incorrect charges. From the ITS technology point of view, this would probably call for more complex systems. Solutions could be provided by various strategies, including a possible shift to more 'vehicle-based' architectures, capable of leaving all data, including the cost calculation, in the hands of the user, or else an extended use of protection procedures to ensure that personal data are not misused. However, most people accept the loss of privacy which comes with mobile phones knowing the precise location of users, so perhaps the 'civil liberties' issue will be more manageable than is often feared. There may be some reluctance to accept itemised billing, which allows family and employers to see one's movements. A smartcard system or one which sends bills only to those who fail to pay the charge might be more acceptable. In Oregon, one option being tested is AVI (Automatic Vehicle Identification) linked with an in-vehicle odometer, which stores information about the road user charges due. The driver then pays the charge at the filling station when stopping for fuel. This may be a route to a charging system which maintains at least discretion, if not privacy (since the AVI has to identify the vehicle's position in order to 'tell' the odometer the correct charge per mile).

5.2.3.5 *The impacts of road user charging*

The impacts of road user charges can include:

- Reduced congestion;

- Reduced environmental damage (less pollution);
- Reduced costs to businesses (faster deliveries due to reduced congestion);
- Revenues raised for investment;
- Better use of the transport infrastructure (higher fees on the most densely used routes leading to a reduction in demand and switching to other modes/routes);
- Social effects (e.g. lower costs in rural areas, increasing accessibility).

Different types of scheme permit different benefits. In schemes such as Trondheim and London, investment funds are raised by adding road user charges to the existing costs of driving. On the other hand, if the aim is to re-balance the costs of motoring on the 'polluter pays' principle (Glaister and Graham, 2003), then the current blanket fuel tax should be removed, and charges redistributed in proportion to the environmental impact (pollution and congestion caused), but with the overall costs of motoring remaining the same. Many motoring organisations are adamant that this is the only acceptable form of widespread road user charging. The development of an approach which will be politically acceptable is crucial and this will require the identification of benefits which are clearly seen to outweigh the disbenefits and a system which is technically sound.

5.2.3.6 *User Acceptance*

One of the difficulties, here, is fear of the new. Road user charging should bring real benefits to drivers in most conditions. Those who do need to drive in the busiest places and at the busiest times will get a better quality of service. Those whose journeys switch to different times or who choose closer destinations will get cheaper motoring. In rural areas, most drivers will get cheaper access. But people generally dislike change. In the UK, there was initially strong opposition to the introduction of variable speed limits on motorways (to produce smoother traffic flows in congested conditions), but by 2002 (Social Research Associates, 2002) many drivers asked for more 'controlled motorways', perceiving time-saving and reliability gains. It may be that the experience of better driving because of road user charging will be the primary means of increasing acceptance (Lehman et al., 2002).

Similar attitudes have been shown in different European countries versus regulations. In Italy, access control to city centres or generalised pricing schemes encountered strong opposition initially, but after a few years there was a strong request for extension. Again, the reason for this change was the tangible benefit obtained from the measures. It is also true that pollution, congestion and safety are today perceived as 'priority problems to be solved', to a degree that any credible method to reduce tangibly the negative impacts could be socially accepted.

5.2.3.7 *Impact on user behaviour*

The aims of road user charging can include encouraging people to travel at different times or to use less congested routes, perhaps choosing different destinations. The integration of RUC systems with information systems is the way to deliver this. With route planning advice in the car and reduced congestion on the road, driving could become a pleasure again.

On the behavioural side, understanding is weak. Ubbels and Verhoef (2003) point out the difficulties in anticipating the behavioural response to road user charging. The value of time may prove important to behavioural change in response to RUC. Drivers are being 'sold' value charging as bringing time savings and reliability improvements – how important are these? Wardman et al. (2002) gives an overview on the value of time, but Gunn (2001) reminds us that people 'place little or no value on time savings or losses under three minutes or so' and 'that unit time losses are valued greater than unit time gains.' Jara-Diaz (2001) develops a distinct method for evaluating leisure time travel savings, which is linked to the ITS and RUC question not least because people may need better information about the costs of journeys they take rarely. Social Research Associates (2002) shows that reliability is valuable.

5.2.4 Future opportunities

The key action needs for the 6th FP relate not only to technical development but also to developing a clear picture of how the technology can serve the distinct policy objectives in different road user charging frameworks. Another critical action is understanding likely user responses. That is important for the road pricing scheme itself, but also critical for knowing what information people will need, where they will need it, how they access it and respond to it in terms of travel behaviour.

Requirements for Research

There are four key areas of general enquiry:

- To review and determine the most appropriate technologies for location, charging and enforcement for general application of charging systems across Europe in all environments, including urban areas;
- To study the best solution for the envisaged 'European Road Charging Service', within an open architecture;
- To assess the socio-economic and environmental effects of road user charging;
- To review the behavioural response to road user charging and the contribution of ITS to making road user charging acceptable to users.

Further actions within the ITS domain:

- Continuation of the research on road user charging technologies and methods, with the aim of exploring the following:
    - The development of a coherent set of quality indicators for the accuracy, integrity, continuity and availability of information technologies for different road user charging schemes with different policy objectives;
    - Implications of interoperability in a free market context of RUC and other services (information, booking, enter-

tainment) in the vehicle and with portable equipment; definition of overall architectures.

In support of the actions decided in the Directive 2004/52/EC:

- Independent technical expert groups to support the 'Electronic Toll Committee' established by the Directive and achieve the link between the research, the projects and such a Committee.

5.3 Road and Traffic Monitoring

Over recent years there has been considerable development of traffic management schemes and emergency scenarios for cities, regions and even trans-national entities to improve traffic flow, reduce congestion and increase performance of the respective road transport networks. For real-time implementation however, many of these plans are hampered by the fact that still little is known of the actual situation on the roads. Experience shows that implementing reliable, automated road traffic monitoring technology usually takes considerable effort. Often little attention is paid to the unassuming task of road monitoring, which provides the very basic functionality for every road traffic telematics application.

Road monitoring is not an end in itself but usually serves road and traffic management schemes, user information in its broadest sense and – as an off-line by-product – provides data for statistics and planning purposes. ISO-1997 defines requirements for traffic monitoring within eight groups of user services:

- Traffic management;
- Traveller information;
- Vehicle systems;
- Commercial vehicles;
- Public transport;
- Emergency management;
- Electronic payment;
- Safety.

For each of these services, the required data and special quality criteria are considered. In most cases several technologies are capable of providing these data, with specific implications on accuracy, costs, scalability or multi-functionality. Some techniques are well established in practice, some are more like pilot applications, and many are expected to be improved in quality and cost.

A system architecture of road traffic monitoring may be described at three levels: the conceptual, the technical and the institutional. The conceptual architecture describes the functions of components or sub-systems of the information chain from data acquisition to information provision to the end-users, distinguishing space and time, and the information flow between those components. Figure 9 shows a rough structure of such an architecture. Possibly, additional decentralised devices can undertake parts of functions of the measurement devices or the traffic information centre. The technical architecture deals with the conversion of the concept into a physical system.

A special problem from the view of engineering is the non-technical issue of institutional arrangements. There are many actors (individuals and institutions) involved in a transport monitoring system:

- Public and private management and information centres;
- Road users evaluating the processed information they get and including it into their decision on driving or travelling actions;
- Road users measuring the driving process and environment conditions (by on-board equipment) or observing special events (especially incidents/accidents) themselves;
- Public and private institutions operating their own measurement systems;
- Professional drivers of public transport, taxi and freight transport fleets, possibly with extended on-board equipment;
- Weather and air quality observation entities.

The diversity of actors implies two problems: Each actor makes his/her own assumptions for checking the correctness of the input data (plausibility) and modelling the calculation of derived parameters. This is ian issue of quality management. At a certain degree, the data exploitation and information processing requires an evaluation of the relevance of data. This needs the determination of preferences which may differ between the acting groups. The problem is more severe in a traffic management system taking decisions than in a monitoring system.

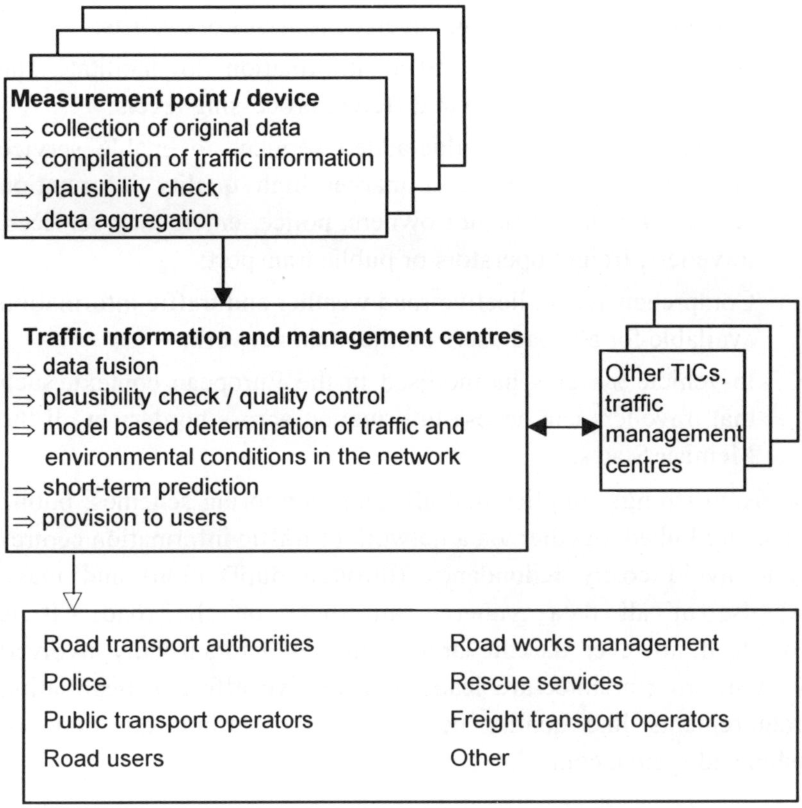

Figure 9: Conceptual architecture of data monitoring

5.3.1 Vision

It is neither possible nor feasible to precisely predict future technological breakthroughs or development threads to formulate a vision of road and traffic monitoring. However, it is possible to identify a number of core elements that seem indispensable for a range of development scenarios:

- Full coverage of primary road network (TERN) with road, weather and traffic monitoring infrastructure;
- Trusted and cost effective technology to reliably monitor all road users, trucks, cars and even pedestrians or cyclists;
- An open market for traffic information to facilitate the exchange of information at different processing levels;
- A viable market for value-added services to enable service providers to generate and market high quality information services for infrastructure owners, police, emergency services, travellers, freight operators or public transport;
- Comprehensive, collective road weather and traffic information available for all road users on the main network;
- In-vehicle systems harmonised in the European context, such that travellers can access information across borders in all the Member States;

Such a vision strongly implies that all road monitoring schemes, public or private, are linked together via a network of traffic information centres (TICs) to avoid costly redundancy (through duplication) and make optimal use of all data gathered anywhere on the roads. It is acknowledged, however, that the large number of actors usually involved in any monitoring architecture leads to extensive effort in negotiating user requirements like quality benchmarks or settling questions of ownership and system control.

5.3.2 State-of-the-art

In most European countries, the past has seen small to medium sized monitoring schemes, usually concentrating on the very primary (motorway) network, conurbation areas with very high traffic loads or specific accident-prone spots. Technology has been proprietary, with even two systems from the same company hardly able to communicate. Even long-lasting efforts at the EC-level to develop a common system architecture have not yet led to interoperability or an extensive exchange of data. Rapid advances in technologies and computing power have left a wide mixture of monitoring devices, communication lines and hardware platforms on the road and in the monitoring centres.

Traffic management and monitoring is more of an issue in densely populated, middle-European countries, and even there, there are major differences between urban and rural areas. Most stationary monitoring schemes are dedicated to the motorway network.

In the Nordic Countries, the weather plays a major role for winter maintenance purposes, even with little traffic on the roads. Adversely, in thinly populated South-Western or South Eastern countries, few monitoring efforts have been undertaken outside the biggest cities.

Besides public authorities, private service providers have started their own monitoring schemes especially in those countries regularly suffering from congestion and delays. Also in some countries operators of toll roads and motorways have set up private monitoring schemes to facilitate operation of their networks.

5.3.2.1 Research programmes

The Telematics Application Programme (TAP) was the major source of funding for research on road monitoring, incident detection and network management in the 4th research Framework Programme (FP4) of the European Commission. The main theme of the TAP was the demonstration, validation and assessment of telematics applications, rather than basic research in new technologies. Generally, budgets for new hardware development or infrastructure implementation have been rather limited, so projects were trying to utilise existing data sources and generate additional value by combining information from several sources and providing software tools for operation and management of these services (information platforms, traffic information centres). Most of the projects focused on urban transport environments, with a multitude of data providers and several transportation modes available.

The successor of the Telematics Application Programme was the Information Society Technologies (IST) programme in 5th FP. Transport Research including intelligent transport systems was funded under the general action line 'Transport and Tourism' within key action I 'Systems and Services for the Citizens'. The focus was even less on basic research, but on practical application and demonstration with a clear market perspective. Only a few projects have been carried out in the area of road traffic monitoring and network management. In the project CARSENSE technologies have been developed that should give sufficient information on the car environment at low speeds in order to allow automated low speed driving. The combination of several sensors will allow improved object detection already based on today's sensors. With the development of new sensor functions such as detection of fixed obstacles and wider field of view, the systems will be capable of being used in urban areas in comfortable ACC (Autonomous Cruise Control) systems.

In the First Call for Proposals of 6th FP which addressed the Strategic Objective 'eSafety and Air Transport', the project ISMAEL aims to determine whether recent advances in magnetic sensors could provide a better means of surface movement surveillance at airports. Therefore physical principles are used as the starting point to provide the best magneto-resistant material for the task. This will be integrated into an advanced sensor head which will form a low cost detector. Within the GST project, the subproject on extended floating car data (XFCD) develops a business plan and a viable architecture to reduce the equipment and communication costs through standardisation efforts. This is considered a pre-condition to achieve the high density of floating cars necessary to achieve the desired monitoring coverage and accuracy on all roads.

The Trans-European intelligent transport systeMs PrOjects (TEMPO) for ITS deployment in the Trans-European Road Network (TERN) are part of the eEurope action plan under the responsibility of DG TREN. Funding is organised via seven Euro-Regional projects for a six-year period (2000-2006). The EC support (16 %) is being spent on the following subject areas:

- The implementation of high quality road monitoring infrastructure for reliable ITS services;
- The establishment of a European network of traffic centres;
- The removal of bottlenecks and easing of traffic flows through traffic management and control measures;
- The deployment of easy access to high quality traveller information services, including the interface with other modes of transport;
- The enhancement of safety and efficiency of freight transport through fleet and freight management systems;
- The development of easy and efficient electronic fee collection systems;
- The promotion of road safety and efficiency through incident and emergency handling;
- Horizontal issues such as systems architecture, standards, data exchange, evaluation, enforcement, organisational and legal issues.

In the proceedings of the TEMPO conferences at Düsseldorf (2003) and Vienna (2004) achievements and future perspectives have been published by the individual projects and specific expert groups by subject area.

Road monitoring is the theme of one of the TEMPO expert groups. The achievements on best practices on monitoring deployment are documented as part of a workshop of its VIKING project (2003). The focus is on the harmonisation via European guidelines for high quality levels of services and for the evaluation of the level of deployment of monitoring infrastructure. In addition, an inventory of the state-of-the-art of existing monitoring systems was given. From the workshop it was concluded that there has been improvement in the harmonisation and development of monitoring toward similar objectives in the long run, in addition to the documentation being useful for experts who will learn about the technological, operational and functional advances in the monitoring domain (Kulmala et al., 2003).

5.3.2.2 Motorways

Road traffic monitoring on busy European motorways has matured from the early, experimental phase and is aiming at a more comprehensive coverage of the whole network. Different approaches pursued by a variety of public agencies and technology providers left a heterogeneous heritage of system architectures, technical solutions and operational procedures. Current stand-alone projects have focused on traffic hot spots and urban motorways, producing a largely incompatible patchwork of monitoring systems in European countries. As major public investments have already been poured into these projects, it is unlikely that existing roadside technology will be replaced by a single superior system, but rather supplemented and integrated into it. The integration of data stemming from a variety of different sources, technologies and content owners will continue to be a major challenge for any provider of comprehensive information for a larger network.

Drivers for the early investments in traffic monitoring technologies have been public agencies interested in road and traffic management (e.g. by variable message signs). Even though monitoring equipment was fixed at the roadside and quite expensive, investments could be justified with public benefits (reduction of accidents, time savings) and sometimes packaged within (even bigger) general construction budgets. Another strategic advantage was the common availability of communication and power lines along the motorways. Monitoring efforts on toll motorways have been pushed for similar reasons. Even on privately owned or operated sections, investments for traffic management facilities could be justified through a gain of customers or though raising the general attractiveness of a toll link. In any case, (contactless) tolling equipment already requires an extensive communication and monitoring input in the first place, sharing costs with corresponding traffic monitoring devices.

5.3.2.3 Secondary network

Justification for monitoring investments on the secondary network and also on the primary network in peripheral countries with low population density is much harder. Often there is little evidence for substantial time savings to be gained by traffic management measures, the general technical configuration is lower, secondary networks are larger and funds lower. As a result, only very few, mostly urban, secondary traffic links have been fitted with fixed roadside monitoring equipment. However, as extensive traffic diversion plans on the public side and real-time traffic information for individual routeing services on the private side become more popular, coverage of all the networks is the challenge. Hence the extension of monitoring schemes onto secondary networks will remain a major issue for traffic monitoring in the future.

5.3.2.4 Private enterprises

Within the last ten years or so private enterprises have entered the market. The idea was that high quality, reliable, real-time traffic information should be a viable service to a growing customer base. In some cases, access to public data sources could be negotiated, though most of the service providers preferred to stay independent from heterogeneous public data sources. As long as venture capital was readily available, large investments were made in independent monitoring networks. Layout could be focused on the specific needs of the traffic information services; full coverage of the motorway network always was a major requirement and concentration on one or two major technologies kept investment prices per unit down and operating costs low. However, despite the very positive prognosis for the market development, aftermarket services have not become viable as yet. The industry is hoping to increase sales by packaging their services with navigation equipment, which is becoming more popular in new cars. The increasing sales of telematics platforms by the car industry also provide the opportunity to recruit larger fleets to generate floating car data (FCD). Meanwhile, private service providers still refrain from investments in secondary network monitoring, even though they know of the pitfalls when diverting their customers off the major motorways.

5.3.2.5 Sensors

Obviously, there are significant technical differences in the public and private approach to road traffic monitoring. Public agencies operate local schemes of fixed, roadside monitoring equipment (mostly inductive loops). CCTV cameras are commonly used for visual incident detection and traffic quality assessment by the traffic control centre staff. Lately, overhead radar or infrared detectors are becoming more popular, often in combination with VMS superstructures. Public monitoring schemes usually take advantage of existing data and power lines along the motorways.

By contrast, private data providers rely on autonomous 'mobile' overhead detector units, usually attached to motorway bridges or lighting poles, powered by solar panels. Communication is wireless (GSM), imposing severe restrictions on data bandwidth for cost reasons. Full national coverage of the primary (motorway) network is usually required for marketing reasons. To achieve better coverage of larger parts of the network most companies work on the generation of floating car data as a cost-effective approach. Private enterprises usually have neither access to live video images nor sufficient staff for visual monitoring. Live traffic web cams leave the visual monitoring to the customer; however, resolution and image frequency are insufficient for the automatic generation of traffic data.

5.3.3 Issues

5.3.3.1 Algorithms

Both public and private data providers claim their respective monitoring technologies are mature and sufficiently accurate for the specified tasks. While this may hold true for the generation of raw data at the roadside level, there are strong indications that new algorithms could better utilise these raw data to generate actual information. Both public and private actors tend to keep data processing secret as there seems to be little readiness to update and improve existing schemes. Updating and improving algorithms in existing data collection schemes could be the cheapest way to produce higher quality road traffic monitoring data.

5.3.3.2 *Quality standards*

Requirements and quality standards for traffic management measures seem to be different from those for traffic information data; as a result there is little co-operation between public and private operators on the exchange of traffic information. Public and private monitoring networks overlap in many places, neglecting the opportunities to utilise additional monitoring cross-sections or to cross-check results. Validation of monitoring data and systematic verification of information is a weak point both in public and private monitoring schemes.

5.3.3.3 *Fault, fraud, mistakes*

Monitoring schemes with large numbers of cross sections and devices, complex communication chains and often 'black box' data processing units provide ample opportunities for fault, fraud and mistakes. Mix-ups of cross-section locations, traffic lanes and travel direction, the malfunction or even complete failure of devices sometimes are hard to detect, yet they can easily devalue the results of a complete system. Future road traffic monitoring technology needs to emphasise even more the inherent auto-monitoring capabilities, at best compensating for human errors in the configuration of systems.

5.3.3.4 *Low cost monitoring*

The technological concepts seem to be capable of producing traffic monitoring raw data at the required level of accuracy. Both public and private operators of equipment argue that high costs (unit prices for fixed or mobile roadside equipment and communication costs in all private schemes) will continue to delay the rapid deployment in the larger parts of the networks as long as traffic information services fail to recover costs to a greater extent. The search is on for new, substantially cheaper monitoring technologies that would allow for the coverage of large, even little-used, parts of the networks. FCD (Floating Car Data) seemed to close the gap at first; however, high communication costs have put a limit on the idea. Recruitment of participants is possible, despite concerns of privacy violation, yet this still requires substantial discounts to be offered on the sale of the essential telematics platform.

5.3.3.5 Piggybacking on existing infrastructure

The overwhelming success of mobile (GSM) phones throughout Europe could provide an opportunity to piggyback on existing infrastructure and surplus data to generate traffic information at very low costs. Opinions on the technological potential are mixed. On the one hand the United States expects telecommunication service providers to develop an adequate technology to locate their customers quite precisely in case of an emergency call; on the other hand there are some strong arguments that data will be too coarse to actually monitor traffic flows, even on motorways. Again, some privacy issues apply which must be solved, and – even a bigger issue – all the data is at the courtesy of the communication service providers who do not seem to see a real business case in developing the technology. Traffic monitoring via mobile phone location could be a promising, cost-effective technology. A clear business case is required to raise interest with the telecommunication service providers.

5.3.3.6 Environmental data

While sensors and concepts for road traffic monitoring (volume counts, speed, stratification) are available and reliable at their respective costs, this does not seem to hold true for collecting stationary environmental data. A key parameter for road safety is the road surface condition, yet there are still no sensors to measure friction directly. Estimation and forecast models calculate slipperiness from precipitation, moisture, surface and temperature of the pavement. However, on a local level and in critical conditions these models are not sufficiently accurate to operate a warning scheme or speed control solely on their results.

5.3.3.7 *Probe data, XFCD*

Some hope comes from the recent development of probe vehicle data. Large sections of the network could be covered accurately that way, but high costs prohibit the equipment of large fleets at the moment. However, the automotive industry is investing considerable effort in the development of vehicle-autonomous sensors for road surface friction to improve vehicle safety (Advanced Driver Assistance Systems, ADAS) at reasonable costs. This extended floating car data (XFCD) could also be communicated to roadside infrastructure to support authorities with collective traffic management and warning systems.

5.3.3.8 *Nano- technologies*

Developments from nano-technology (the capability of building things one molecule at a time) could provide a totally new detection structure for monitoring how traffic is flowing. Originated by the MOTES/DUST/SMART DUST approach of the University of California at Berkeley, the technology makes use of low-energy sensors and communication units. The new sensors will be able to detect vibration, footsteps, voices and still images and transmit them to a network of fixed and mobile relay collection stations. The significant reduction of size will enable sensors to be deeply embedded in the physical world products or materials and spread throughout the environment like smart grains of sand. Currently, sensor networks with communications capabilities have been produced that are as small as a penny. In the future, nano-technology will create miniature sensors so small they could be woven maybe even within the ink on a piece of paper (Pister, 2004).

5.3.4 ***Future opportunities***

Although many road traffic monitoring systems exist in a lot of countries, it is obvious that there is demand for improvements. Further technical developments will support this demand. Regardless of which scenario of a future monitoring system will become reality, certain problems in all three sectors of concepts, technologies and institutional arrangements need particular attention. They are summarised below.

At conceptual level, the following tasks need improved solutions:

- Fusion of traffic data from various sources (local measurement points, floating car data, video observations, manual input) to describe traffic situations and in particular to detect incidents;
- Short-term and medium-term traffic prediction based on historical and current data as well as on models, considering the anticipation of expected route changes caused by current information, and control strategies and their acceptance by the drivers;
- Estimation of dynamic origin-destination traffic streams;
- Road surface condition assessment and forecasting based on historical and current data as well as on models.

Technical problems exist in the following areas:

- Low-cost measuring and communicating floating car data;
- Video image processing, especially under adverse visibility conditions;
- Measurement of road surface conditions (wetness, black ice).

Finally, institutional issues are:

- The co-operation of data owners;
- The establishment of public-private partnerships by adequate contracts, clarifying the competences of involved actors, and the financing.

In addition, the problem of quality concerning the equipment level required for the various partial systems is of relevance. In particular, the minimum number of vehicles required to deliver floating car data seems to be very promising within the scope of an integrated monitoring system.

The following table (Table 6) gives a detailed overview of the issues of road traffic monitoring concerning different services or systems; they may serve to formulate specific actions for future opportunities in development and implementation.

Table 6: Road traffic monitoring

Basic road transport system information	
Concepts	Standardisation of network and traffic description
Traffic management	
Concepts	Control system requirements of: UTC (traffic lights, parking, etc.), access control/ramp metering, speed limits/warnings, incident detection, routeing (community/user optimisation, O-D), short-term traffic forecast, road weather forecast Data fusion, sources: private traffic, public transport, freight transport, weather stations, special non-automated information
Technologies	Local traffic sensors: inductive, optical, acoustical Positioning/movement (time/space): GSM/UMTS, map matching, video image processing, aerial view, satellite (GPS) Data transmission: FM/DAB (radio), DSRC, GSM/UMTS, satellite Information display (optical, acoustical): control centre, roadside, in-vehicle
Institutional	Co-operation of data owners: PPP contracts, data exchange, info/control strategies, liability, financing
Traveller information	
Concepts	Requirements of: routeing (community/user optimisation, O-D), traffic conditions, weather conditions Data fusion: (→ traffic management)
Technologies	Local sensors: (→ traffic management) Data transmission: (→ traffic management) Information display (optical, acoustical): stationary terminal, roadside, in-vehicle
Institutional	(→ traffic management)
Incident/emergency detection and management	
Concepts	Classification of incidents/emergency cases
Technologies	Mobile data transmission
Single vehicle operations	
Concepts	Requirements of control criteria, assistance or automated driving, strategies for cases of malfunction, special infrastructure for automated traffic
Technologies	Sensors: for vehicle movement and of neighbouring vehicles, environment Data transmission: vehicle-vehicle, vehicle-infrastructure
Institutional	Legal issues: traffic law, product liability, privacy protection

Road pricing	
Concepts	Network determination, closed/open system, pricing criteria, integration with transport system planning
Technologies	Fee collection: GPS/digital map/GSM, DSRC - enforcement techniques
Institutional	Legal conditions, financing, system operation
Incident / emergency detection and management	
Concepts	Classification of incidents/emergency cases
Technologies	Mobile data transmission
Environment surveillance	
Concepts	Prediction and assessment of road conditions
Technologies	Road surface (ice, slipperiness)
Commercial vehicle operation	
Concepts	Requirements of special surveillance and routeing
Technologies	Weighing in motion
Public transport	
Technologies	Travel demand acquisition: data transmission, stationary, mobile
Safety services	
Concepts	Public travel security, safety for vulnerable road users, intelligent junctions
Technologies	Mobile communication, vulnerable road user detectors, pedestrian detectors
Winter maintenance	
Concepts	Modelling of road surface condition from environmental data
Technologies	Cost-effective vehicle road friction sensor, cloudiness sensor

Many issues of concern for a substantial improvement of road traffic monitoring have been identified in the previous section. Institutional arrangements should be addressed in a political context, yet quite a number of technical subjects emerge as a starting point for further research activities. The basic concern is that current technology is too expensive for covering large parts, yet alone all, of our road networks. While technical solutions do exist more or less to cover every monitoring desire, actual implementation concentrates on accident hot spots and some heavily used parts of the motorway network. For the systematic diffusion of monitoring schemes, technology needs to become more cost-efficient.

There are four main directions for research:

- Cheaper sensing and communication technologies for stationary observation to facilitate more widespread use of traditional (usually public) concepts
- Reducing costs for XFCD-technology, telematics platforms and vehicle-to-road communication to allow the operation of larger monitoring fleets, covering virtually every part of the network;
- Exploring the spin-off effects of nano-technologies for low energy detection and communication for totally new traffic and incident detection schemes;
- Improving algorithms and models for the interpretation of existing data to generate higher value information at the lowest prices.

These points hold true for both traffic monitoring and road environment surveillance, with the exception of road friction sensing. Here, accuracy and reliability of sensors and models still need real improvement to support road safety management measures.

Future opportunities could be organised along the following project lines:

- Road surface condition monitoring and prediction;
- Cost-effective monitoring technology including nano-technologies for service quality;
- Piggybacking on cellular phone data – cheap ways to area-wide traffic monitoring;
- Area-wide mid- and long-term traffic forecast and prediction;
- Satellite and airborne image processing for road traffic and driving environment monitoring;
- Develop public-private partnerships based on business plans and open architectures to promote data fusion of all levels of transport networks and sensing technologies.

These opportunities have to be seen within the tasks and objectives of traffic management. Their fulfilment will support and close the loop to these tasks and are a means for the take-up of the technologies envisaged for future transport management schemes. The development of these architectures and the need for standardisation have to be addressed for multimodal and intermodal requirements and other developments of traffic information and control systems.

Chapter 6

Freight Transport

The efficient transport of freight is fundamental for industry and business, and a major factor in the smooth working of the economy. However, the high percentage of freight carried by road in Europe - over 90 % in some countries – makes it a major cause, and victim, of traffic congestion. This is compromising the efficiency of transport operations, and also has serious implications for safety and the environment due to accidents, vehicle emissions and noise.

Over the last two decades, the European Commission has made efforts to 'rebalance' the modes by promoting the use of non-road transport and encouraging intermodal solutions. For a number of reasons, this has not proved easy to achieve. One fundamental difficulty is that about 80 % of goods movements in Europe involve trips of under 200 km. Intermodal transport is at present economically viable only for distances of at least 400 km. The outcome is that road haulage is in most circumstances the most competitive solution for shippers in terms of direct costs, in spite of the growing risk of delays caused by road congestion.

Given the forecasts of continued growth in freight volumes - an estimated annual rise of between 2.0 % and 2.7 % - pressure to contain the negative impact of goods movement and to find commercially valid alternatives to road haulage is bound to intensify. One of the strategies being proposed is to embrace more fully the potential offered by ITS. Telematics applications can certainly make an important contribution to the efficiency of freight transport, but it is argued that their success will depend on an appropriate regulatory framework for transport, as well as the willingness to undertake strategic investments in the supporting infrastructure.

This chapter attempts to identify the aspects of freight transport where telematics systems could make the most effective contribution, and the areas where research and other actions are required to make their wide-scale deployment a real possibility. The first part looks at goods transport as a whole, especially longer distance freight movements, and examines in particular the potential of telematics for supporting intermodality. The second part considers the specific problem of goods deliveries in urban areas. It should be emphasised, however, that these are not totally different worlds; in many cases they represent different stages of the same freight journey, and the ITS applications involved are similar.

6.1 Long Distance Freight

6.1.1 Policy background

Traditionally, the policy of the European Commission has been to promote 'sustainable' transport; this concept being understood as an approach which balances environmental, social and safety factors, with the achievement of economic efficiency (cost-effectiveness for both the providers and users of transport services) and high service quality. The search for ways of reconciling these often conflicting objectives has underlain the Common Transport Policy for many years.

The document which sets out the current strategy is 'European Transport Policy for 2010: Time to Decide' (European Commission, 2001). Many of the sixty guidelines relate specifically to freight, and underline the importance of revitalising the railways, promoting the transport of goods by inland waterways and short sea routes. An updated version of this statement is currently being drawn up.

One of the difficulties is to favour non-road transport without imposing excessively restrictive measures or heavy additional costs on haulage operations which could affect Europe's economic competitiveness. The long-term goal is to create conditions in which alternative solutions can become commercially self-sustaining.

6.1.2 General trends in freight transport

Among the trends affecting freight transport in Europe are:

- The globalisation of production, which is changing the patterns of freight distribution (e.g. the growth in intercontinental sea freight is increasing the importance of ports as nodal points) and intensifying the level of competition, accentuating time and cost sensitivity. The overall effect is to strengthen demands to reduce the 'frictions' in goods transport;
- The outsourcing of transport and logistics operations resulting from the increasingly specialised transport requirements. This is leading to strong consolidation in some sectors, such as parcels, which are now dominated by a few large international shippers;
- The growing use of e-commerce practices, which affects the organisation of the supply chain, tending to favour 'just-in-time' production/delivery and the fragmentation of deliveries (i.e. smaller and more frequent shipments);
- The demand for flexibility, due to rapidly changing demands for transport and also the problem of unexpected delays (e.g. caused by accidents or traffic congestion). This is leading to the practice of 'mobile warehousing' where routeing and load distribution is rescheduled while freight is on the move, and makes it essential for fleet managers to have access in real time to the information necessary to be able to choose the best routes, services, or even modes, and recalculate arrival times;
- The introduction of increasingly stringent regulations regarding certain products, such as fresh food and hazardous goods, which will require tracking of consignments.

One of the consequences of these changes is that shippers are no longer expecting a 'transport only' service, but are seeking complete logistics solutions with transport operations that can be fully controlled (i.e. the possibility of monitoring vehicles and the cargo/container throughout the journey). The ultimate target is a service that offers 'zero delays, zero inventory, and zero red tape'.

Progress in this direction can be facilitated by ITS applications. By offering logistics 'packages' providing high quality end-to-end services, and enabling data exchange between modes, telematics tools can make freight transport more efficient and also favour intermodality.

Within this wider (and more ambitious) vision of an ITS-enabled freight transport system, the efficient movement of goods will be an important element in the strategy for improving the competitiveness of business.

6.1.3 Vision

The future foreseen for the freight domain is one in which ITS will be part of the 'armoury' of all the players involved: the freight shippers and transporters, infrastructure managers (i.e. those responsible for running the transport networks and related facilities, such as freight villages or terminals), as well as public authorities (who use telematics tools to monitor and control freight transport operations, e.g. road use charging, access control and enforcement). Together, the integrated use of these tools goes to constitute an 'Intelligent Freight Transport System'.

The widespread use of fleet management platforms helps road haulage operators to improve productivity through the ability to plan and monitor missions with greater efficiency, and achieve better use of vehicle capacity (through load consolidation). Continuous tracking and communications between vehicles and the Fleet Control Centre will make it possible to reschedule operations while trucks are 'on the move' and adapt to changing requirements. Access to real-time information on traffic, weather and road conditions will make it easier to produce accurate arrival time estimates and avoid congested parts of the road network.

Drivers will benefit from a range of services, accessible through onboard fixed units and handheld devices, and operable in any European country. These will provide them with support, especially in unfamiliar areas (e.g. location of service areas, hotels, etc) as well as greater safety for themselves (e.g. automated emergency service alerts in the case of an accident) and their loads (e.g. automatic tracking and temperature monitoring). Preventive safety will be substantially improved by the use of co-operative systems which enable communication between vehicles in a local area, and also between vehicles and the infrastructure. This will make it possible to receive warnings of dangers caused for example by an obstacle or accident on the road ahead, icy patches, fog banks, etc. in time to take avoiding or corrective action.

In this vision of the future, *intermodal (combined) transport* will be used far more widely than at present. The ability to exchange documentation between vehicle and control centre before arriving in a freight terminal, to track consignments across modes, and automate the 'processing' of loads and vehicles in terminals will significantly improve efficiency. The effect will be a reduction in the time loss in transfer operations as well as better service quality and security. The consequent increase in the use of rail, inland waterways and short sea routes as part of intermodal journeys will help these to reach a critical mass, triggering a virtuous circle of higher productivity, lower tariffs and increased business.

There will exist a series of long distance 'freight ways' extending across Europe N-S and E-W operating regular road/rail shuttle services. Similar opportunities will be exploited using inland waterways and maritime cabotage. The efficiency of transferring the freight will be facilitated by improvements in the physical infrastructure and equipment. Terminals sited at key nodes will offer facilities for bi- or tri-modal transfers.

The freight sector will remain firmly in the private domain, stimulated by open competition between operators. Common standards are nevertheless adopted, so systems are interoperable. Public-private partnerships (PPPs) help to promote the adoption of ITS in critical areas such as cross-border services. Co-operative arrangements exist between Trade Associations and private firms, especially Small and Medium Enterprises (SMEs), operating in the transport sector and give the latter the benefit of access to advanced management tools and ITS platforms.

6.1.4 State-of-the-art

6.1.4.1 Road freight

An overwhelming majority of the inland (road, rail and inland waterways) freight in European countries is carried by road - the average is around 75 % (by tonne-kilometres), while in some countries it is over 90 %. Despite the growing frequency of serious traffic congestion, the relative convenience (direct door-to-door service) and cost advantages of road haulage have until now made it difficult for other modes to substantially increase their share. An exception in the overall picture for non-road transport is the substantial growth in maritime container traffic in recent years.

While road haulage is far more efficient than in the past, in terms of logistics as well as reduced energy consumption and toxic emissions, it is nevertheless still characterised by:

- A multitude of uncoordinated transport operations, especially in the many small-scale fleet operations;
- A high percentage of unused load capacity (due in particular to empty return trips);
- A substantial amount of time 'wasted' in various types of delay, as shown in Figure 10.

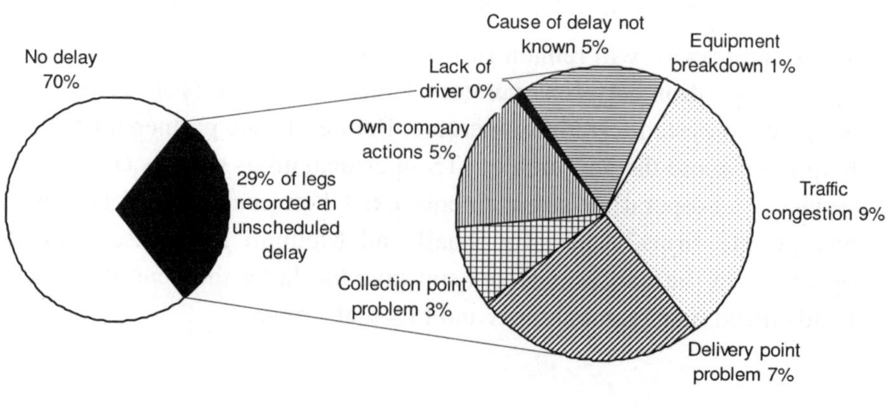

Figure 10: Causes of delay in food deliveries (McKinnon and Ge, 2003)

Although road haulage generally offers the most rapid mode of delivery for shippers, as indicated by Figure 10, almost a third of trips are subject to unscheduled delays. The two principal causes are problems at the collection or delivery point and traffic congestion. This picture is confirmed by a survey carried out by Eurolog (2003) which indicates that for much of the time vehicles are in fact 'non-productive', i.e. idle or waiting (see Figure 11), and suggests that considerable scope still exists for improvements in efficiency.

Telematics tools can make a positive contribution in the majority of these areas of inefficiency. These tools include:

(a) Automated scheduling systems which are able to calculate the most efficient sequence of deliveries and loading/unloading operations. For large firms with multi-drop deliveries, it is estimated that they permit an increase in the productivity of personnel of around 25 % (Léonardi and Baumgartner, 2004), as well as substantial fuel savings by improving the load factor and reducing overall mileage;

(b) Fleet management platforms, which use satellite technologies for tracking the location of vehicles. They offer a range of functions which can help the transport manager in both monitoring trips and in the planning stage by producing detailed logs of the missions undertaken (e.g. fuel consumption, stops made, distances travelled etc) and calculating statistics. Such platforms can also be integrated with onboard sensors to offer, for instance, anti-theft systems and vehicle diagnostics, which facilitate the planning of maintenance operations.

(c) An interface between the control centre driver through a fixed onboard unit or a PDA (mobile handheld device) permitting the exchange of messages and also automation of administrative processes, such as the confirmation of goods delivery. They can also provide added-value services for the driver including route guidance and 'infomobility' services

Automated fleet management platforms are now beginning to penetrate the European freight transport market. In a sample survey carried out in

Germany in 2003 (Léonardi and Baumgartner, 2004), around 17 % of the trucks were equipped with on board units. In many other countries, and especially among small firms, the percentage is likely to be much lower. It is undoubtedly a market with great potential for growth.

An indication of what can be achieved through substantial investment in ITS and the adoption of e-commerce practices is at present offered by the major freight shippers, especially integrated carriers such as DHL, UPC and TNT, and the principal food retailers. Having the advantage of large-scale operations, a single company environment, and activities covering the full logistics chain, the extensive use of telematics tools enables them to gain very significant efficiency benefits. The challenge is to spread this level of efficiency to the transport community as a whole.

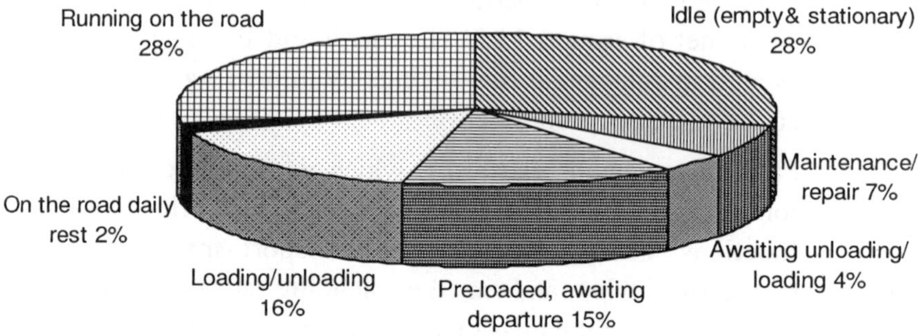

Figure 11: Average vehicle utilisation over 48 hours (McKinnon and Ge, 2003)

New developments, for which research is already underway, include the so-called ‘co-operative systems’ whose objective is to improve traffic efficiency and safety. These are advanced telematics applications which will permit communication between vehicles in local ‘ad hoc’ networks and also between vehicles and the infrastructure. Other initiatives involve the improvement of the quality and accuracy of driver information services with the use of 3D maps and ‘floating car’ data.

6.1.4.2 Non-road freight and intermodality

After a period of notable growth in the 1990s, the demand for intermodal transport seems to have reached a plateau. Since 2000, it has levelled out at about 5 % of total tonnage. The question is whether this represents a maximum viable ceiling or whether the difficulties currently experienced in encouraging shippers to use intermodal solutions can be overcome.

Behind the average statistics regarding the modal split lie considerable national variations. In Switzerland, for example, only 36 % of freight travels by road, and in Sweden 42 %. By contrast in Spain and Ireland, road haulage accounts for well over 90 % of inland goods transport.

The reasons for these differences are partly geographical and partly the result of organisational and regulatory factors. Certain physical features, such as the mountainous terrain, the existence of seaports, rail nodes, and inland waterways can favour the use of rail or waterborne transport. But there is also evidence that well designed regulations, incentives and organisational framework can have a beneficial impact. This means that it is important to understand how these can be supported by ITS.

The figures quoted here are drawn from a survey by the Observatory on Transport Policies and Strategies in Europe (OTPSE, 2005) carried out for the French Conseil National des Transports (CNT).

Intermodal freight has a much higher penetration in the road-rail sector (25 %) than maritime (10 %) or river traffic (5 %). While 20 % of road-rail freight adopts the so-called 'piggy back' solution (combined transport), 80 % is unaccompanied (mainly containers and swap bodies). Intermodal transport is predominantly in the hands of a few large shippers and is not easily accessible to smaller players. Given the current growth in maritime traffic and air traffic, there would appear to be potential for intermodal terminals located at seaports and airports.

The countries with the highest tonnage of road-rail freight are Germany, Italy, Switzerland, France and Austria. While in the first two, the volume of combined transport is currently growing, in the others it is shrinking. In Austria, the decline has been attributed to removal of the 'Eco-points' incentive; whereas in France it is generally put down to improvements in the motorway network which have favoured road haulage. This illustrates the difficulty in permanently capturing the market and in making intermodality pay without financial incentives. As well as higher direct costs, rail and waterborne transport also suffer from a lower service quality than road haulage in relation to reliability, punctuality and security.

Intermodal transport is technically and economically viable only in a very specific market segment: the large scale movement of non time-sensitive goods over a long distance (the threshold is estimated to be around 400-500 km). It is ideal for mass transfer of freight between two distant points involving a route which presents practical difficulties for road transport, such as trans-Alpine routes.

Intermodal solutions in Europe are in fact concentrated on major axes, such as the N-S corridor between the Rhine delta and the North of Italy, passing through the Alps. The most successful and efficient approach has proved to be the institution of regular shuttle services on key routes.

Telematics systems are as yet not widely used for intermodal transport except for the road leg of the journey and specific functions, e.g. freight terminal access. The constraints are both organisational and technical, including the large number of players involved and the difficulty of achieving interoperability between modes and across national borders.

Among the ways in which ITS can support intermodal transport are by:

- Improving efficiency (time-saving) in transfer operations, e.g. by permitting data exchange between freight terminals and vehicles on the move before their arrival;
- Improving the quality of services offered (e.g. tools to facilitate the tracking of containers across modes, anti-theft systems for the rail/waterborne legs) and their continuity across borders;
- Making it easier for small transport operators to gain access to intermodal services;
- Providing a more accurate expected time of arrival (ETA) (for many cargoes the critical factor is not so much the speed of transport, but the precision of the arrival time prediction).

6.1.5 Issues

The first key issue regarding the use of ITS for road haulage is to identify and promote a 'killer application' or trigger which would lead to onboard units being universally installed on vehicles.

The higher the level of market penetration, the greater the benefits both for individual firms and for the community as a whole. As indicated above, fleet management platforms offer transport operators the possibility of substantial gains in efficiency and productivity. In addition, the cumulative effect of load consolidation and better routeing at European level would result in lower overall mileage for the volume of goods transported and lower environmental impact. Furthermore, when a high percentage of vehicles are using advanced telematics applications, such as 'co-operative systems', there are further benefits in both safety and traffic efficiency.

So how can fleet managers be persuaded to invest in such systems? Clearly on-going reductions in cost will help, as well as more widespread awareness of the positive impact on efficiency. But to accelerate their adoption, the missing factor is a strong enough incentive, or mandatory requirement, for the installation of onboard units to become widespread.

There is however a further condition. In order to gain the full advantages described above, it is necessary for such platforms to be interoperable. If systems are developed independently using incompatible approaches, then many valuable benefits could be lost. To establish services which are able to exchange information with other systems and will be operative in any European country, requires a common architecture. A second critical issue is therefore how to guarantee the interoperability of services between types of system and across national borders.

The opportunities for providing these two issues is discussed in section 6.1.6.

The third issue regards the thorny question of intermodality. As freight transport is a highly competitive sector, shippers will inevitably opt for the mode which offers the convenience, good service and low prices. This means that costs, time and reliability are critical factors. If left to market forces, it seems likely that, at least in the immediate future, the share of non-road freight will not expand significantly. So what are the prospects for the future?

Given that the volume of freight seems destined to continue to grow, the long term future seems likely to offer a scenario which could be highly damaging for the European economy. If current trends are to continue with no constraints on road haulage and no significant investment in alternatives (such as rail freight), congestion will gradually intensify, delays will become commonplace, and pollution and safety will become even more serious issues than at present. At some stage, the road will no longer offer a commercially feasible option for goods movement on many routes and the market will be forced to find other solutions. Since the environmental lobbies would probably block the construction of new roads, investment will begin in creating alternative networks, but since the building of new infrastructure (e.g. rail routes) is a very long term undertaking, freight transport would remain highly inefficient until such work was completed. The only way to avoid such a situation would be to foment an 'intermodal shift' before the critical point is reached.

One of the strategies proposed by the environmental lobby to accelerate such a shift is to 'internalise' the external costs of transport. However, a study carried out by the European project RECORDIT indicates that the comparative cost advantage of using intermodal transport is slight, as shown in Figure 12. This suggests that even if shippers were obliged to pay the indirect costs of transport, the impact might not be sufficient to lead to a significant variation in the modal split.

It is of course possible in the longer term that changes in the underlying conditions, such as a substantial increase in fuel prices, could tip the balance more decisively. Nevertheless, in the meanwhile, the only really effective approach would be to adopt a 'dual strategy' consisting of:

(a) Strong financial incentives or regulatory measures to constrain road freight, such as the banning of Heavy Goods Vehicles on certain routes or a Europe-wide introduction of heavy tariffs for trucks;

(b) A concerted effort to render intermodal transport more efficient and attractive in its own right.

The RECORDIT project in fact concludes that it is perfectly feasible to achieve substantial cost reductions in intermodal transport by:

- Undertaking radical reform of the organisation of intermodal transport, with the objective of drastically reducing the costs of handling, infrastructure and the rail part of the journey;
- Promoting a co-operative effort by the many players involved;
- Focusing such efforts and resources on a given number of long distance axes which offer the most favourable opportunities.

Telematics applications are fundamental to the operation of an automated tolling system and also in making the organisation of intermodal transport more efficient. Examples are given in section 6.1.7.

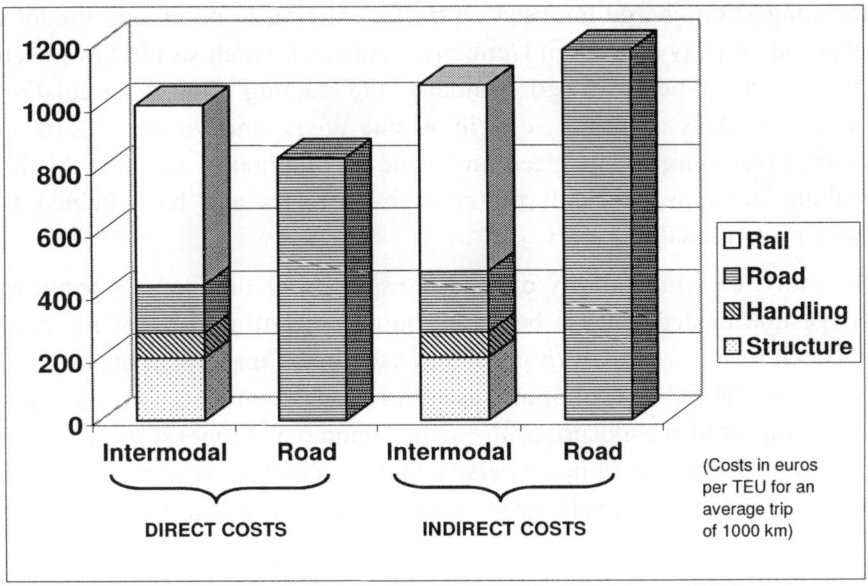

Figure 12: Comparison of costs: intermodal vs. all road freight (RECORDIT project)

6.1.6 Future opportunities

6.1.6.1 Transport policy in relation to intermodality

The changing context, changing attitudes, and advances in technology are setting the scene for a more optimistic outlook. Pressure from environmental groups for containing the volume of road transport is intensifying. There is growing awareness of the external costs incurred by road accidents, pollution, noise and congestion. This is reflected in increasingly stringent emissions and noise standards as well as safety targets which are obliging the authorities to take more decisive action in regulating vehicles and traffic circulation.

The change of attitude is evident in the acceptance of schemes such as the congestion charge imposed on traffic entering London, and the tolls imposed on heavy trucks in Germany, neither of which would have been conceivable some years ago. Similarly, the banning of HGV circulation on certain days or routes, e.g. in Alpine areas, and greater efforts to enforce regulations (e.g. speed limits and driving hours) are undoubtedly making life more difficult for haulage operators and have helped to favour intermodal transport.

Secondly, the vulnerability of road transport to disruption as a result of congestion or accidents is becoming more evident as parts of the road network reach saturation (a problem exacerbated on E-W routes by the entry into the EC of central and eastern Europe countries). One example is the impact of the closure of the Mont Blanc road tunnel resulting from the fire in 1999. For shippers persistent delays and uncertainty regarding arrival times can eventually erode the advantage of road transport.

6.1.6.2 Interoperability

If ITS applications are to be really effective they need to be integrated, and for this to happen they must be interoperable. To ensure in turn the interoperability of ITS applications, there must be agreed standards and a 'high level architecture'. This implies a common approach to the planning of the telematics systems. Compliance with such a methodology and common standards guarantees the possibility of data exchange between different types of system, even when developed independently (for a more detailed discussion see section 7.1).

Fortunately there already exists a European ITS Framework Architecture developed by the EC project KAREN in the late 1990s and extended by the FRAME projects (2001-4). Although this architecture covers mainly road-related ITS applications, interfaces to other modes are included. The architecture has adopted elements relating to road haulage proposed by the COMETA project and also the Italian national ITS architecture ARTIST. More specific requirements and recommendations for freight across all modes were examined by the THEMIS project. For short sea shipping and inland waterways similar initiatives have been carried out and a KAREN-related architecture is now available.

An excellent opportunity therefore exists to make use of this basic framework and to build on it further in order to extend the European ITS Framework Architecture to cover the new and emerging needs of freight and fleet management systems.

6.1.6.3 Trigger for equipment of vehicles with onboard units

The key to the universal equipment of fleets could possibly be a European road pricing strategy for freight vehicles. If such a policy were introduced, the onboard equipment required could provide the necessary platform for a wide range of ITS services and could underpin a move towards more intelligent transport logistics. A system of differentiated tariffs for use of the road infrastructure would require interoperability throughout Europe.

In order to achieve interoperability of the software and hardware systems involved, the European Community can play a valuable role, not only by promoting common standards and common operational procedures, but also by supporting co-operative initiatives.

There are already examples in the transport industry where competing companies (especially SMEs) have initiated community systems for their mutual benefit. In some instances this has been done as a defence mechanism to protect members from competition from large integrated carriers, in others it is in response to the need to meet a legislative requirement, which has acted as a catalyst for setting up broadly spread systems permitting all players to comply with the regulations.

Among the constraints to the use of automated fleet management systems are the lack of specialised personnel and the organisational impact (Léonardi and Baumgartner, 2004). This is a problem felt in particular by small transport firms. One way for them to benefit from advanced freight transport platforms without the need to set up an in-house Operations Centre is to use web-based fleet management functions. These are open platforms which are offered on a service-contract basis and can be accessed via a normal PC. As well as avoiding heavy initial investment, the solution offers flexibility, as the choice of functions can be tailored to meet individual needs as they evolve.

A useful role for the EC would therefore be to launch a virtuous circle with policies which encourage 'networking' between different transport modes, and providing cost incentives for co-operation. Such incentives should not become subsidies for unprofitable services, but should focus on infrastructure and other up-front investment for subsequent services.

The European Community could also support trial schemes based on perceived real requirements. The initial requirements could be defined by the Trade Associations but, in order to ensure successful implementation, these would have to work with private sector organisations. The onboard platforms now being provided by OEMs can be good starting points for intelligent freight transport systems.

Dissemination is also essential for improving the knowledge of possible applications of ITS and making the benefits made more widely known amongst transport operators. This could be achieved by making available details of selected 'best practice' cases.

6.1.7 Recommendations

6.1.7.1 *Role of the EC in promoting ITS for freight transport*

Among the actions which could be promoted by the EC are:

- Extension of the European ITS Framework Architecture to deal in greater detail with freight transport. This could possibly be done by setting up of a Task Force consisting of representatives of European countries which already have national ITS architectures covering freight, as well as initiatives such as the Euro-regional projects with an interest in facilitating long distance (cross-border) goods traffic;
- Promotion of initiatives with the objective of encouraging networking between transport firms, the sharing of information platforms;
- Promotion of open web-based fleet management platforms which are easy for SMEs to access on a service contract basis (as outlined in the previous section).

6.1.7.2 *Actions to support intermodal freight*

To favour intermodality, actions need to focus not only on promoting the use of telematics tools for transport operations, but also innovative systems to support management and business processes.

One of the key objectives of research would be to develop an 'electronic market place' making it easier for transport shippers and operators, especially SMEs, to gain accessibility to intermodal networks, even in the more peripheral regions of Europe. Such initiatives could involve:

- The promotion of on-line 'freight service centres', i.e. web-based information platforms, able to provide potential customers with all the necessary information – tariffs, booking, documentation requirements – for organising an international freight journey;

- Electronic Freight Planners offering a demand-responsive service built on real-time information and able to support load matching for intermodal transport;
- The pooling and bundling of freight transport demand along given corridors to help reach the critical mass which would make intermodal transport services economically viable;
- The drawing up of international rail accessibility standards and development of advanced systems for rail slot management and pricing for intermodal train services on national railway networks (with the aim of abolishing cross-border frictions);
- Development of tracking systems to allow consignments to be traced through the whole journey, assuring better security and also accurate arrival time estimates.

Intermodal operations as 'connection points'

- Evolution in concept of intermodal terminals: from modal 'break points' to modal 'connection points' able to guarantee the smooth transfer of loads and efficient processing of information and documentation;
- Promotion of best practice: rather than 'reinventing the wheel', opportunities should be used to learn from operational experience in existing automated container handling terminals. Examples of innovative solutions for integrated intermodal transport chains should be documented and made known.

Total quality management (TQM)

One of the important objectives for intermodal transport, in order to increase its acceptance, is to improve the level of service quality. This requires the development of TQM systems which can guarantee quality standards, especially for:

- *Punctuality and reliability*: when shippers have to deliver within a very tight time margin (or risk paying fines), they require support in route planning and changes to adapt to actual road conditions. This can be provided by tracking and tracing systems which, together with real-time traffic information, can produce constantly updated arrival time estimates (ETA). To guarantee the service levels required by customers, such tools must be integrated with advanced management systems (including transport incident detection, assessment and interception strategies). (On the other hand, when cargoes are not time-sensitive, shippers need to focus on the accuracy of the ETA rather than reduced journey time. This requires a different approach);

– *Security*: the challenge is to provide seamless surveillance and control systems throughout the whole transport chain and across international borders;
– *Monitoring of sensitive cargoes*: for safety and health reasons it is essential to know the exact location of hazardous goods, perishable foods or special cargoes, such as animals).

Such systems need to be integrated within a global TQM concept, which includes all stages of the transport process (from planning to booking, dispatch, delivery and invoicing) and all parties (operators, transfer terminals, etc).

Improved logistics integration

Operators of intermodal transport should be able to share information with the supply and distribution chain management of their customers. This would help them to incorporate intermodal options into their logistics planning at an early stage and would also enable intermodal service providers to prepare and operate their services more efficiently.

Further specific telematics issues that require research, refinement and/or demonstration are the following:

– Ways of undertaking intermodal 'asset tracking' of cargoes and vehicles to ensure continuous stock visibility. Strategies are also needed for linking such systems across national borders;
– Harmonisation of message and document standards (to allow the use of a single 'electronic window' for intermodal transport planning;

- Development of e-commerce techniques for freight logistics in conjunction with advanced mobile communications (GPRS, UMTS) and mobile computing instruments (PDAs, GSM-Scanners) with a view to offering web-based added-value services.

6.1.7.3 *Complementary research recommendations*

In addition to research regarding ITS tools, the following areas have been identified for providing important complementary support:

Transport Policy

- A detailed examination of the responsibilities of the players involved in freight transport, e.g. shippers, infrastructure managers, public authorities, etc., with a view to clarifying their roles, eliminating conflict, and promoting better co-operation;
- Analysis of resource conflicts between freight and passenger transport regarding the transport infrastructure;
- Development of benchmarking methodologies for intermodal transport to permit the systematic assessment of strategies;
- Establishment of efficiency criteria for intermodal terminals (bi-modal and tri-modal nodes) in relation to location, level of automation, technical and commercial viability;
- Proposed bundles of measures which could improve the competitiveness of intermodality in freight markets, including financial and regulatory measures as well as incentives to adopt new technologies;
- Development of a Europe-wide open standard for automated tracking and tracing systems (to help counteract the perceived lack of security in non-road based modes, particularly rail).

Transport Economics

- Gathering of statistics for all 25 European member states on intermodal transport with respect to freight origin/destination, type of cargo, volume, value of shipments, etc., and ways of making them accessible to transport planners and policy makers;

- Investigation of the potential of SMEs for short distance procurement and distribution with regard to intermodal transport systems;
- Development of a methodology for evaluating different approaches to infrastructure use charging for freight transport (e.g. road and congestion pricing);
- Examination of ways of increasing the economic viability of intermodal nodes and transfer points by providing value-added logistics services.

6.2 Urban Deliveries

Among the particular features characterising the delivery of goods in urban areas are the facts that:

- Environmental and safety factors are especially critical, both in shopping districts and residential areas (pollution and noise levels as well as the safety of other road users and pedestrians);
- Delivery vehicles in cities face strong competition with private cars and public transport for road space as well as bays for loading/unloading;
- Constraints are frequently imposed on vehicle circulation by the nature of the road network, e.g. narrow streets, or by regulations, e.g. pedestrian precincts and areas in which motorised access is restricted.

Though goods deliveries represent a relatively small part of urban traffic (around 10-30 %), the impact on congestion and the generation of toxic emissions is disproportionately high due to the type of vehicles used (large/medium trucks) and the fact that they need to park in busy streets while making deliveries.

Efforts are therefore being made, especially in the larger urban areas, to reduce the nuisance this causes to other users of the city. In addition, many city centres have been 'redesigned' to increase their attractiveness to shoppers and visitors. But these initiatives, which frequently involve the creation of traffic free zones, can make life very difficult for the shippers as well as the businesses they serve. Pressure is therefore mounting from both sides to find satisfactory ways of managing urban logistics.

A whole range of measures are currently operated in European cities, but the principal strategies involve:

- Instituting access restrictions in specific parts of the city;
- Imposing limits on parking/loading, and more recently;
- Making monetary charges for use of urban roads.

These measures are operated with numerous variations, e.g. restrictions on certain size or type of vehicle, 'time windows' for loading/unloading operations, charges for parking, entry into certain parts of the city depending on the emissions status of the vehicle, and so on. The problem is that these regulations are frequently not only complex and confusing for delivery operators, but also difficult for the authorities to monitor and enforce fairly. Schemes which are not well conceived invite non-compliance (drivers may prefer to park close to their destination and take the risk of a fine, rather than use an official unloading area!).

ITS instruments can offer valuable support in managing such regulations without the need for a large personnel dedicated to the task. Considerable potential exists for the development of specialised telematics tools, both in the management of the deliveries themselves, and in helping the city authorities to regulate more effectively the movement of commercial vehicles in urban areas.

6.2.1 *Vision*

In our future vision for urban freight deliveries, co-operative agreements will have been reached in all major European cities between retailers, shippers and the local authorities on the access arrangements for commercial vehicles, especially to city centres. These will often involve the definition of strict 'time windows' and the need to book (and in some cases pay) for authorised access, but at the same time offer the advantage of using reserved lanes and designated (un)loading areas or transhipment points.

The result is a substantial reduction in the 'nuisance' to general traffic, while permitting the efficient servicing of business needs. It is facilitated by the use of specially designed ITS tools which support the management of the more complex strategies, and make the system transparent.

From the point of view of the telematics systems and equipment used, operators have moved away from isolated proprietary solutions towards integrated multi-purpose, multi-user platforms, which take full advantage of the opportunities offered by high-precision satellite location (using Galileo services) and mobile computing. An important feature of these is the possibility of access to web-based supporting services and information (traffic, maps, routeing, booking of delivery slots, etc.).

A driver approaching an unfamiliar city can request details of the local access arrangements before entering the urban area, download a city map onto his onboard unit and request a suggested route to his destination. The 3D map shows him landmarks to facilitate route finding, indicating streets which are currently congested or too narrow for his vehicle to use. He has requested advance authorisation to enter the restricted city centre zone; the necessary fee is paid automatically and registered by his firm's fleet management platform. As the vehicle has EURO-4 emissions status, it is allowed to enter this area.

Another vehicle belongs to the fleet of a large logistics group which has a sophisticated system for optimising its deliveries and collection. The drivers know the area well so they do not need navigation support, but the transport company finds it very useful to have information on typical traffic flows on the network in order to plan their operations. They can also benefit from access to the city's mobility platform to be able to book the use of one of the transhipment points and arrange a special electric load-carrier for the last stage of delivery in a pedestrian precinct.

A third driver works for a small local delivery firm which has equipped its vehicles with inexpensive onboard units. The transport manager uses a central terminal (PC) through which he has access to a web-based fleet support service that enables him to plan the order of 'drops' and monitor the fleet. The firm takes advantage of co-operative arrangements with other small companies, which allow them to share vehicles when there are non-urgent consignments to make to a common destination. In this case the loads are left in a special container at a transhipment point.

Deliveries in city centres are facilitated by the existence of 'urban hubs' where loads can be transferred to energy efficient, low noise, low emissions carriers designed for deliveries in urban areas (e.g. rapid and quiet loading). The items are packed in containers which are equipped with RFID (Radio Frequency Identification Devices) tags to enable automatic recognition with handheld computers (PDAs) and constant tracking of the delivery status.

The typical architecture of the kind of open mobility platforms which make this possible is shown in Figure 13.

By means of a SOAP (Simple Object Access Protocol) interface, the city mobility authorities and the fleets themselves have access to a series of web-based services. In this way, the authorities can make available information on access regulations or facilities that they offer. If vehicles are equipped with GPS antenna, then their presence in restricted areas can be monitored and the information used to support enforcement measures against unauthorised vehicles. Vehicles with onboard units can also exchange messages with the Control Centre or City Authority terminal, in order for example, to request authorisation to enter a restricted area, or to check in real-time whether the official loading bays are free.

For deliveries in residential areas, a variety of solutions are used, including service points sited in strategic locations (e.g. locker-boxes for collection/deposits at service stations, car parks, and within residential areas). Customers have access to unmanned points by means of electronic keys which function with personal codes. The operations carried out are registered online and printed confirmation with legal validity given.

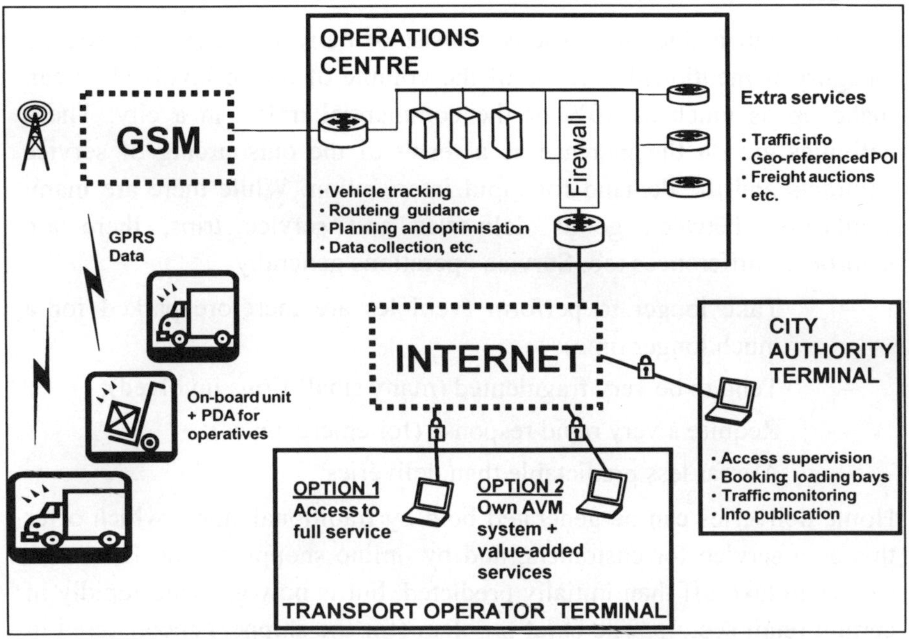

Figure 13: Open platform for the management of urban deliveries

6.2.2 *State-of-the-art*

In most cities the volume of delivery traffic is growing. There has been a marked increase in particular in the demand for transport of small parcels and non-containerised goods. Turnover in the sector has been rising by 8 % per year since 1996.

Commercial traffic in urban areas falls into three main categories:

(a) Deliveries to retail outlets;

(b) Service and maintenance activities;

(c) Home deliveries.

The first category is highly diversified, its features depending on the type of business being served. This will determine the size of loads involved, the frequency of delivery and any special conditions (e.g. need for refrigeration), which need to be taken into consideration

Even though it does not strictly consist of delivery traffic, the second category is mentioned because of the volume of traffic involved. It can make up as much as 30% of the commercial traffic in a city. These activities are on the increase as a result of the outsourcing of service contracts and the demand for rapid intervention. While there are many similarities between goods deliveries and service trips, there are significant differences too. Service operations generally:

- Take longer to perform (vehicles are therefore parked for a much longer time);
- Tend to be very fragmented (many small firms involved);
- Require a very rapid response (for emergencies);
- Are far less predictable than deliveries.

Home deliveries can be generated both by traditional stores which offer this as a service for customers, and by online shopping. The latter was slower to take off than initially predicted, but is now growing rapidly in certain retail sectors. The chief problem (for the shipper) encountered in this form of delivery is that of finding someone at home to accept the goods. From the environmental and safety point of view, there is the problem of increased traffic circulating in residential areas.

Although local authorities are far more likely than in the past to adopt specific strategies for delivery traffic, it is not easy to find a satisfactory approach. The general tendency is to apply restrictive measures such as access controls (which can be put into practice at short notice without the need to set up any special facilities), but these can lead to dissatisfaction and difficulties on the part of transport operators and the businesses they serve. An approach being considered in some cities is that of setting up points where loads are transferred to smaller vehicles for delivery to the customer. They need to take into account however that any ‘break point’ in a delivery trip will add to the overall time and costs. Efficient logistics are a fundamental element in the cost strategy of retailers.

The adoption of e-commerce practices in general (both B2B and B2C) is leading to the emergence of new logistics patterns. While the traditional model for retail distribution is represented by (2) in Figure 14 below, it is increasingly common to find variants such as (1), where goods are delivered directly to the final customer (home deliveries) and (3), in which 'order fulfilment centres' are set up specifically to deal with the despatch of orders taken online. These often serve a whole region, but sometimes also have local depots in the periphery of the urban area where goods are transferred to smaller vans for delivery directly to the 'doorstep'.

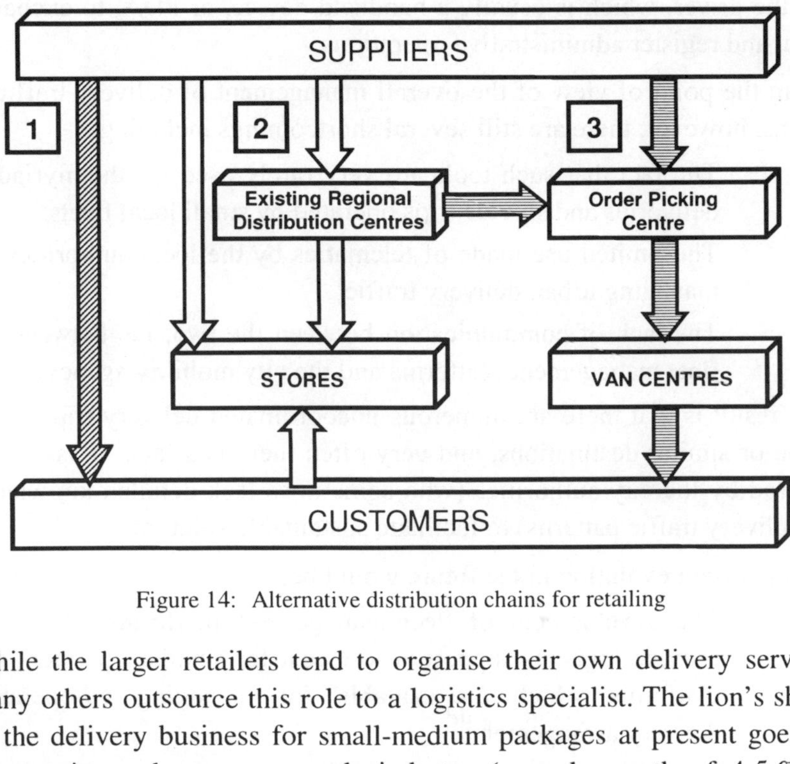

Figure 14: Alternative distribution chains for retailing

While the larger retailers tend to organise their own delivery service, many others outsource this role to a logistics specialist. The lion's share of the delivery business for small-medium packages at present goes to the courier and express parcels industry (annual growth of 4-5 % is expected in this sector over the next five years).

The problem faced by all is that of 'the last mile'. Various strategies are being adopted to reduce the number of unsuccessful trips caused by not finding customers at home to take delivery. One alternative to doorstep deliveries which has been successfully piloted in several countries is the 'pickpoint'. These may be manned or automated service stations, such as the 'packstation' deployed by Deutsche Post/DHL in Germany and the Chronospare system in France. The items themselves can be identified by the operator's tracking and tracing system in order to confirm delivery and notify the receiver of an item awaiting collection.

Telematics instruments are beginning to play a more prominent role in the management of goods deliveries in cities. As in the case of interurban freight transport, the larger-scale operators seek to optimise the logistics (and reduce costs) by using fleet management platforms with an interface for the driver, which is usually a handheld device, or PDA, to exchange data, and register administrative operations.

From the point of view of the overall management of delivery traffic in towns, however, there are still several shortcomings including:

- The fact that such tools are very rarely used for the myriad of deliveries and service trips operated by small local fleets;
- The limited use made of telematics by the local authorities for managing urban delivery traffic;
- The lack of communication between the two, i.e. between the fleet management platforms and the city mobility agency.

The result is that there are numerous uncoordinated delivery trips to the same or similar destinations, and very often there is a lack of assistance offered by the city authorities (who sometimes lack detailed knowledge of delivery traffic patterns) to facilitate sustainable solutions.

An important evolution in the future would be:

- The development of fleet management platforms which can provide logistics support at low cost for smaller operators and platforms which are suitable for managing co-operative delivery arrangements;
- The development of telematics tools able to help local authorities in monitoring, charging and controlling delivery traffic, in providing services and also making available information to delivery operators.

6.2.3 Issues

6.2.3.1 Strategies for Regulating Urban Deliveries

The majority of European cities have now adopted specific strategies for delivery traffic, some of these developed as part of European research and demonstration projects. City authorities have at their disposal a wide range of approaches, tools and measures for regulating the movement of freight in urban areas. It is not always easy for them, however, to choose the best solution from the many alternative policies and technologies available.

In order to make a sound choice it is necessary not only to have good knowledge of the various possibilities, but also an accurate picture of the situation regarding freight delivery in the city concerned. This requires:

(a) The availability of clear guidelines based on best practice and results of demonstration projects, including not only technological aspects but also the management implications;

(b) Adequate data to enable the patterns of freight movement to be better understood. While detailed (but usually confidential) data is collected by private shipping companies to improve the efficiency of their operations, limited information on delivery and service traffic is in general available to public authorities.

The automated collection of data is a function which could be integrated with the traffic monitoring function. It is important to be able to find ways of doing so efficiently (so the costs are low), in a practical way (so the data collected is really useful) and without violating data privacy rights.

6.2.3.2 Co-operative approaches

One of the concepts put forward to optimise urban deliveries, especially for the smaller operators, is 'Collaborative Management'. This involves co-operation between different carriers, with the aim of sharing vehicle space for consignments which have a common destination. There are conflicting views on its potential. Certainly it is not easy to organise due to the difficulty of meeting the needs of numerous individual enterprises operating in a competitive environment, to problems of data confidentiality, and the overheads of handling and planning which increase operational costs. Attempts to use this approach have often overlooked the fact that deliveries are part of long supply chains, and that city logistics cannot be viewed in isolation from the rest of the supporting network.

Some successful examples exist, but at present the level of optimisation is low compared with that obtained by professional operators. It seems nevertheless likely that the voluntary exchange of information between firms requiring transport services will in future become more common, especially in places where restrictive conditions are imposed on vehicle access.

Collaborative approaches are particularly appropriate for deliveries in city centres, for example in managing the use of low-emissions vehicles for the consignment of goods in pedestrian areas. The only practical way of managing such operations is by means of open telematics platforms to which all interested parties have access. These would be used as a form of electronic 'market place', assigning goods to the available vehicles in an optimum fashion and tracing operations from a central terminal.

6.2.3.3 Open telematics platforms to support efficient logistics

Numerous systems offering fleet management functions already exist on the market. An evolution of interest for small scale delivery operators, and possibly also the management of collaborative initiatives, concerns the creation of 'open' platforms, which allow access at any time to a large number of parties while preserving confidentiality of the data exchanged.

The most practical way of providing such a service is through web-based platforms with protected access. Some small scale and partial examples exist, but otherwise this is an area in which pilot schemes and demonstrations are required.

The use of open platforms has potential for several functions relating to urban deliveries:

- Collaborative approaches to freight distribution, especially in relation to the management of urban transhipment points;
- Value-added services regarding e.g. real-time traffic information, mapping and detailed road network information;
- The publishing of information by city authorities regarding current access regulations in given zones;
- Offer of services by city authorities, such as booking of delivery slots, occupation of loading bays, access authorisations, etc.

One of the main technical challenges regards the successful integration of the value-added services. From the business viewpoint, such platforms would have the function of offering value-added services to firms, but leaving the market free to find its own user equilibrium.

6.2.3.4 *Tracking and tracing of containers and individual items*

One of the critical requirements of the advanced management of urban deliveries, especially if co-operative approaches are adopted, is an efficient way of tracking individual items or the containers in which they are packed. A solution already being used by some of the larger shippers is that of RFID (Radio Frequency Identification Devices). RFID tags (or transponders), which can be active or passive, are attached to the goods. The passive tags used today can be read from a distance of about two metres and store just enough data for an identifier and a small amount of other information. This data can be stored, displayed and/or sent to another location. Active RFID tags have a greater reading range (about 15 metres), but are larger and cost more. While passive tags can last indefinitely, active ones last only as long as their battery.

In urban applications, RFID tags are valuable for tracking items which pass through various stages of transhipment. The advantage is that the tags can operate in any kind of environment and do not need to be visible to the reader. Passive RFID tags can also be used to map and identify urban containers to be loaded onto the vehicle in order to speed up the operation. The use of active RFID tags on delivery vans or trucks could, for example, make it possible to check that only authorised vehicles are using a given loading-unloading place. If attached to containers or single items, they can provide data relating to goods themselves.

Over the next few years, experimentation involving both types of RFID technology by the larger freight shippers is likely to continue. This will no doubt determine the most appropriate applications of the different types of tags and possibly also produce further innovations. If the cost of active tags were to fall substantially in the future, this could open the way to their widespread adoption for urban deliveries and facilitate the management of schemes involving the transhipment of goods.

It is desirable, however, for standards to be established at European level so that such tags are interoperable and can therefore be an integral part of co-operative initiatives as well as for operations by individual firms.

6.2.3.5 Platforms for city mobility authorities

There are three main types of activity which city mobility authorities are likely to require increasingly in the future, and which involve telematics applications:

(a) The control and monitoring of delivery traffic, including the imposition of access restrictions and charging for the use of road space;

(b) The publication of information in order to make known current rules and regulations, the zones affected, the vehicle categories permitted, the services and facilities offered, such as information on availability of special parking areas;

(c) The provision of services for delivery or service vehicles such as the booking of loading bays and the use of transhipment platforms, access authorisations, etc.

The development of a platform in which these different functions can be efficiently operated and are also integrated (in the sense that data can be exchanged between them) is a major challenge for future research.

6.2.4 *Recommendations*

While it is the responsibility of the city authorities to draw up appropriate regulations and measures regarding urban deliveries, the preparation of guidelines based on best practice and field experiments at European level would provide a very useful contribution.

An important role for the EC is therefore in the proposal of guidelines regarding the ITS tools available to support the management of goods delivery and the related organisational aspects. These should include a suggested management framework, as well as ways of overcoming the current barriers to co-operation, i.e. how to bring together the many different stakeholders, public and private.

The suggestions for research and other actions are summarised below:

Telematics applications

- Development of a 'reference platform' for city authorities to use for the management and control of delivery traffic and provision of services. It would serve as a basic framework to be built upon and adapted to provide the particular services required in a given urban area.
- Investigation of innovative approaches to the (multi)use of urban space and the development of telematics applications to manage it in a more flexible way. This could include a methodology for optimal space allocation and also ways of organising the flexible assignment of space, e.g. through real-time information on the availability of loading bays, automated allocation of time-slots, etc.
- Proposal of methods of automated data collection to provide information for planning delivery traffic management strategies in a way that is inexpensive, easy to organise and respects data protection laws.

Traffic management policies

- Evaluation of alternative delivery traffic management schemes and coercive actions to minimise impact on congestion, noise, pollution and general nuisance (assessment of the impact of priority lanes, delivery time restrictions, road pricing schemes, emissions/vehicle size limits, planning controls on location of distribution centres, etc.) with the aim of producing practical and quantified guidelines.

Harmonisation and standardisation

- Proposal for the standardisation of the interfaces between on-board tour planning systems and traffic management systems to ensure that traffic information services can be integrated. Standard performance indicators could also be developed.
- Standardisation regarding the size and shape of goods containers to facilitate the use of interoperable handling equipment.
- Examination of norms relating to the storage and handling of perishable foodstuffs at transhipment areas and 'pickpoints' to ensure health and safety standards. Also of monitoring techniques to ensure the rules are respected.
- Investigation of feasibility of introducing European norms relating to operating licenses for delivery vehicles, as well as legislation on the size, weight or emissions of vehicles used in certain areas (e.g. city centres) and special measures such as taxation for empty running, with the aim of reducing noise and environmental impact.

Specific recommendations regarding e-commerce

- For a better understanding of the potential impact on traffic and cities, simulation models are needed:
  (a) To analyse the logistics relating to different product types;
  (b) Evaluate alternative delivery systems to identify the arrangements which minimise traffic impacts.
- Analysis of the behavioural impact on the use of online shopping on individual mobility, including 'trip chaining' and 'replacement trips', transport modes used, etc.

– Research to establish how to favour use of online shopping and home delivery services by sectors of the population with special needs, e.g. the elderly or physically handicapped.

Chapter 7

ITS Support

This chapter examines a set of issues which are fundamental to the future development and uptake of ITS products and services. They have been brought together in this chapter as they all provide support that is not limited to specific ITS sectors but is felt 'across the board'.

The first regards the concept of high level ITS architectures whose aim is to provide a common framework and methodology for the planning of ITS applications. As argued here, the widespread (voluntary) compliance with a European-level architecture can help to achieve the harmonisation of ITS with a positive impact on the ability to establish services which are interoperable across modes, across borders, and across different types of device, and also on industry's willingness to invest in ITS.

Secondly, we examine some of the critical issues regarding positioning. Precise and reliable location is essential to many telematics applications, which means that it is important to understand the implications of new developments regarding radio navigation technologies and the satellite-based services to be made available by the GALILEO initiative.

Finally, we take a look at the 'human element', in other words the people responsible for the design, development, maintenance and implementation of ITS. It is suggested that a co-ordinated approach at the European level to training and education of ITS personnel could significantly improve the skills and knowledge available, and consequently have a positive impact on the quality of ITS applications.

7.1 Architecture

7.1.1 Background

One of the reasons claimed for Europe's success in gaining an early lead in the development of many telematics technologies was the focus on well-defined 'building blocks' such as traffic control, driver information, public transport management and route guidance systems (European Commission, 2000 (1)). These benefited from the numerous European R&D programmes carried out from the late 1980s onwards, which produced promising results for a whole range of ITS systems and services.

By the mid 1990s, however, it was becoming clear that the real-life implementation of these systems was much slower than had been hoped. To address the problem and identify the causes, a High Level Group was set up. Among the conclusions drawn by this group was that ITS schemes too often consisted of stand-alone solutions created for specific applications. If ITS were to achieve its full potential and wide-scale deployment, the systems would need to be interoperable. One of the challenges to be faced was therefore how to integrate these building blocks and to favour interoperability.

The answer proposed was the development of a pan-European architecture that would permit some degree of compatibility and synergy across ITS applications. Compliance with such an architecture would ensure that ITS projects were well conceived from a technical and organisational point of view, and would be harmonised throughout the whole of Europe. This would benefit industry as well as the public sector and end users. Among the recommendations of the High Level Group was therefore the creation of a European Telematics System Architecture for the road sector. This was endorsed in June 1997 by the Transport Council.

The result was the EC-funded project KAREN (Keystone Architecture Required for European Networks) which ran from April 1998 to October 2000. This had the task of developing a European architecture for road-based ITS applications. It was not the only such initiative - architectures were developed for specific ITS sectors, for example by the QUARTET projects for urban traffic control, COMETA for freight transport, and GERDIEN for motorways - but it was undoubtedly the most ambitious.

Creation of the European ITS Framework Architecture

In 2000, the KAREN project published the first version of the European ITS Framework Architecture (European Commission, 2000 (2)). In order to respect the principle of subsidiarity, this consisted of a high level framework or 'model' which European countries could use, if they wished, as a basis for their own national ITS architectures. These could, however, be adapted to reflect local requirements. A further important feature was that the architecture was not technology specific. Its purpose was to give guidelines regarding the content and, equally important, the methodological approach to ITS planning. A somewhat similar approach was adopted in the USA, where the National Architecture (U.S. Department of Transportation: National ITS Architecture), as a federal product funded by the Department of Transport, supports individual States in creating their own architectures. Other non-European countries which have developed national ITS architectures include Japan, Korea, Canada and Australia.

After the conclusion of the KAREN project, two follow-up projects, FRAME-S and FRAME-NET (2001-2004) had the task of encouraging and supporting the development of national ITS architectures. A series of updates have also been made to the original Framework Architecture and two new versions published in 2004.

A prominent aspect of all these projects has been efforts to raise awareness of the need for and benefits of ITS architectures (European Commission, 2004) and the creation of commitment to their development.

7.1.2 Vision

Our future vision consists of a situation where ITS is far more widely deployed in Europe than at present. Readiness to promote ITS on the part of public authorities has been stimulated by clear evidence of the benefits with respect to transport efficiency, safety, and intermodality. Use of ITS services by individual travellers has grown as a result of the practical support they offer for all types of journey. This has all led to an expanding market for ITS developers, manufacturers, suppliers and service providers.

A fundamental feature of this picture, and one of the reasons for the success of ITS in gaining a mass market, is the harmonisation (and hence interoperability of applications) achieved across Europe, across modes and across devices. This has been possible due to willing co-operation between all the stakeholders. These support, and take part in, a continuous 'harmonisation' process, which starts at the ITS policy level and leads in a coherent way to decisions regarding implementation.

ITS deployment is being shaped by a user-oriented view, rather than a 'technology push'. This is possible since all of the main players - local authorities, industry and the final users - are involved in the process. In general the public administrations (Transport Ministries) are leading the way, and are able to influence the direction of ITS policy in their countries, helping to ensure an approach which balances social with commercial benefits.

User involvement is stimulated by the evident advantages being gained through this harmonised approach to ITS planning:

- *Travellers in Europe* (the final users of ITS) obtain benefits from interoperability, especially for cross-border travel and multimodal trips, due to the existence of 'seamless' information systems and services. They also reap the advantage of lower prices, since the extended European market and greater competition reduces the cost of systems, components and services;
- *Public administrations* are able to speed up the development of ITS architectures by using the Framework Architecture as a 'model'. This allows them to achieve earlier and better planned deployment of ITS. They can also benefit from the experience

of others because of the existence of a common 'language' and methodology, and access to examples of best practice;

- *The ITS industry in Europe* has the advantage of an extended market for products, systems and services, since these are guaranteed interoperability throughout Europe. They also have a more stable basis for forward plans and investment decisions.

High level discussions and examination of the implications of European transport policies for ITS lead to voluntary international co-operation to achieve interoperability in key areas of ITS (on the lines of the process which led to the Europe-wide adoption of RDS/TMC). This favours a forward-looking approach to ITS development.

At the technical level, the European ITS Framework Architecture serves as a basis for planning and provides a common methodological approach for describing the possible deployment options. The Architecture itself has been extended to cover interfaces with non-road modes, and is gradually including new ITS areas where harmonisation is beneficial.

7.1.3 State-of-the-art

Thanks to the various initiatives sponsored by the European Commission (especially the KAREN and FRAME projects) a European ITS Framework Architecture now exists, and an increasing number of countries in Europe have already developed, or are now in the process of developing, a national ITS architecture. These include the ACTIF project in France, ARTIST in Italy, TTS-A in Austria and TEAM in the Czech Republic which are all based on the Framework Architecture. Among countries that have expressed interest in following this lead are Spain, Slovenia, Hungary, Romania, Slovakia, Sweden, and the UK.

There are also several independent developments, e.g. in the Netherlands, Finland and Norway, some of which were initiated before the conclusion of the KAREN project. In the first two cases, moves are being made to achieve gradual convergence with the Framework Architecture.

A number of EU regional projects or organisations, such as VIKING and CONNECT, are also active in the architecture area. Further EC projects which developed ITS architectures include THEMIS for freight (including non-road modes), and COMPRIS for inland shipping.

It is clear therefore that, although still not universally supported, there is growing acceptance of the need for national ITS architectures and the benefits of compatibility with the Framework Architecture. While the majority of Europe's public administrations are favourable, and architectures are beginning to be drawn up at city level or for specific ITS areas, e.g. travel information systems, reticence is still expressed by industry - no doubt due to fears of the imposition of standards or other constraints.

The private sector is, nevertheless, involved in efforts to seek European agreement relating to lower level 'technical' architectures in several areas, e.g. safety applications and rescue services (within the GST project), and the development of the so-called 'co-operative systems' which combine onboard and infrastructure-based ITS.

So, apart from the continued need for outreach activities, what more, if anything, is required in the way of action or research to further the use and development of high level architectures at the European level?

Experience, for example in the United States, has shown that it is not sufficient for a national or framework architecture to exist. If it is to be widely used and to remain relevant, its users must be supported and the architecture itself constantly maintained and updated. To continue to respond to real needs, it has to be constantly expanded to cover new types of ITS applications. In this respect it is essential not to lose the impetus created by the KAREN and FRAME projects (nor to waste the sizeable investment made by the European Commission).

The Functional Areas currently covered by the European ITS Framework Architecture are listed in Figure 15. Some possible future extensions have already been identified. Among them are the requirements (user needs) relating to ITS applications for intermodal freight, road user charging, and co-operative systems for road transport, as well as a methodology for the development of business models and the definition of Institutional or Organisational Architectures.

European ITS Framework Architecture: Functional Areas

- Electronic Payment Facilities
- Safety and Emergency Facilities
- Traffic Management
- Public Transport Operations
- Advanced Driver Assistance Systems
- Traveller Journey Assistance
- Support for Law Enforcement
- Freight and Fleet Operations

Figure 15: ITS functions included in the European ITS Framework Architecture

Several of these extensions are in fact already being developed within national architecture projects. A constructive strategy would therefore be to find a way of incorporating them within the Framework Architecture, in order to make them available to other countries. This would require adaptation to ensure they are appropriate for use at European level, but would avoid the repetition of similar work elsewhere.

In relation to the present situation, it is possible to draw the following conclusions:

(a) There is growing acceptance of the usefulness of the European ITS Framework Architecture – at the very least as a tool for helping dissemination and outreach among decision-makers and supporting deployment. In some cases (e.g. France, Italy), the use of the Architecture has gone hand-in-hand with expansion of ITS services; in others it has given valuable support as a basis for drawing up ITS deployment plans.

(b) The Framework Architecture is a permanently evolving 'tool': new ITS services need to be covered, interfaces to other (non-road) transport modes gradually included, and regional needs accounted for. This calls for the direct and 'deeper' involvement of stakeholders (Member States, local and regional authorities, operators, etc) with willingness to contribute their experience. In parallel, this requires technical support in order to maintain some control over methodological and formal aspects. To be of benefit to others, any new extensions or enhancements introduced must, for example, continue to respect the common 'European' methodology and language.

(c) Now that the European ITS Framework Architecture has reached a relatively mature stage, the work of a predominantly 'technical' nature carried out so far (development of a consistent methodology and of the Framework Architecture itself) should be accompanied by activities at the policy level.

In other words, the time is ripe for the development of what could be called an 'ITS Policy Architecture'. This would have the function of ensuring that the technical architecture work takes place within a coherent policy framework. It would also as a result help to establish the priorities for extensions to the Framework Architecture. Sound scenarios for policy choices (and, where possible, policy agreements) are the first and essential steps in the creation of ITS Architectures.

It should be noted that a similar shift of attention is taking place in other parts of the world (e.g. in the USA) where architecture development is at an advanced phase of development.

In the future, it will therefore be necessary to find ways of ensuring that:

- The European ITS Framework Architecture continues to evolve and remain in touch with future needs;
- Expert support is available for countries which wish to develop an ITS architecture or maintain an existing one;
- A mechanism is established to permit the development of a 'policy architecture' or framework.

7.1.4 Issues

The objectives identified above raise two principal issues. The first concerns the way in which the architecture-related requirements of Member States can be met in the future; the second regards the strategy adopted to foster harmonisation.

7.1.4.1 Meeting the future needs of Member States

In relation to ITS Architecture, the needs of European Member States (as well as other groups or projects) fall into three main categories:

- Better understanding of the costs and benefits of ITS architectures for those considering their development;
- Support in actually creating new ITS architectures;
- Support in maintaining an existing (national) architecture.

The first involves all the various 'outreach' activities designed to clarify the nature and purpose of an ITS architecture, and the implications of developing one. From the discussion above, it is evident that such efforts will in the future need to be aimed more widely, addressed not so much to national administrations as to regional and city authorities, specific ITS sectors and, in particular, industry.

The second concerns the considerable number of Member States which have already expressed an interest in developing a national architecture. These would obviously benefit from the kind of support already received by others, such as workshops, training sessions, and a technical Help Desk. Since ITS deployment in many of the countries concerned (e.g. in Eastern Europe) is still at an early stage, assistance in creating a national architecture with European compliance is especially valuable.

The last is necessary because, although a country may already have a national architecture based on the Framework Architecture, it will at some stage wish to create extensions or make variations to meet specific local needs. If this is done without direct reference to the European level, it could result over time in a gradual 'drifting apart', to the point where international compliance is compromised. This could eventually threaten the interoperability of ITS applications.

Until recently, all of these functions have been carried out by EC funded projects. The critical question for the future is whether the maintenance and promotion of the Framework Architecture, as a 'mature' product, should be expected to be self-financing. It could be argued that if users are sufficiently convinced of the benefits, they should be prepared to pay for the technical support they require. On the other hand, there is a risk that smaller 'recent accession' states could find it difficult to do so, and that not all countries would be willing to finance support given to others.

It would therefore seem advantageous for certain types of assistance - such as technical support for new architectures - to continue to receive EC funding, while alternative sources are sought for others. At the time of writing there has already been a move in this direction. Several Member States have grouped together to create a 'Forum' and made a financial contribution for support in ensuring compliancy of the new versions of their architectures, making extensions to the Framework Architecture, and general services, such as website management.

7.1.4.2 Harmonisation strategy: enforced or voluntary?

The growing deployment of ITS, and especially its penetration of B2C markets (such as the use by the travelling public of navigation services, travel information, etc.) is likely to strengthen and accelerate the demand for harmonisation. This is because an important requirement of the customer is an integrated service, i.e. seamless information for travel in different countries, for different modes, and when using different types of device - 'smart' handheld phones, onboard units, PCs, etc.

To achieve harmonisation there are two basic alternatives: it can either be enforced or voluntary. In other words, it can be achieved by means of coercion (e.g. through the issue of Directives), or the willing co-operation of Member States and the industrial players involved. The latter is clearly preferable, but requires a good understanding of the issues involved and the benefits of a co-operative approach, as well as a readiness to take a long term view of ITS development. This is the function of a 'Policy Architecture' or Framework.

7.1.5 Future opportunities

One of the most important opportunities for the future lies in the constitution of a permanent consultation platform with the role of encouraging co-operation between European countries with regard to ITS deployment. It is felt that such a platform would need to operate on two closely linked levels - one strategic and one technical.

At the strategic level, the existence of a consultation body - consisting of high level representatives of the public administrations of European Member States - would offer the chance for open discussion of the implications for ITS of general transport goals (such as increased road safety, intermodality). Possible alternative scenarios and priorities for ITS deployment could be explored.

One of the principal objectives would be to identify the potential for common deployment initiatives (voluntary agreements between Member States or involving the EC), and establish priorities for in-depth studies to determine how best to achieve harmonisation. This, in turn, would provide the basis (and help set the priorities) for extension and enhancement of the Framework Architecture.

The result of this strategy - the interoperability of ITS systems - would help to stimulate market growth by reducing the development and implementation costs for Member States and industry, and consequently reducing costs for the user.

The process described above would be greatly facilitated by the existence of an ITS 'Policy Architecture' or Framework. This would provide a way of describing ITS-related policy choices already taken (or being planned) by Member States. It would state a country's basic mission or philosophy regarding ITS implementation in terms of policies or strategic goals. And by establishing an agreed process (and language) for expressing such a statement, it would facilitate communication and also permit a systematic assessment of different scenarios and their implications.

There is as yet no formal methodology for doing so, but it is suggested that the process could include the following:

- A description of the overall mission for ITS;
- The legal and organisational structures in place;
- A set of policy statements presenting the ITS-related political choices already taken and those planned;

- A definition of the basic requirements for each policy (which would also permit a consistency check);
- The identification of the functionalities required.

To help draw up the Policy Architecture, the High Level Group would need to be supported by an Expert Panel or 'Advisory Group' with the required knowledge and experience of ITS and the world of transport.

A strong link would have to be established to a second level concerned with the technical implications (including the development of standards, protocols, etc.) of the common deployment initiatives proposed by the Forum. This link could consist of a Technical Committee, comprising representatives of National ITS Architecture teams, other architecture-related initiatives (e.g. EU-Regional projects), and possibly national ITS representatives participating as observers.

The most obvious tool for supporting this activity is in fact the European ITS Framework Architecture. For support and guidance in extending or updating the Framework Architecture or developing new national architectures, expert technical advice would be needed. One possibility is to set up a 'Back Office', consisting of a small group of architecture experts whose assistance would be available on request to development teams. They would help to ensure that new architectures respect the common (European) methodology and language. This activity could be supported financially either by European funding or by the Member States themselves.

The latter would be a preferable solution, especially in the long term, as it would mean that the users themselves would act as 'custodians' of the Framework Architecture. The activities themselves could be steered by a pan-European group or Forum made up of representatives of national and other ITS architecture projects. This would initiate a self-sustaining process in which the evident benefits of compliance result in a readiness to invest in its regular upgrading as well as the development of new ITS architectures.

The overall structure is represented in the diagram below (Figure 16).

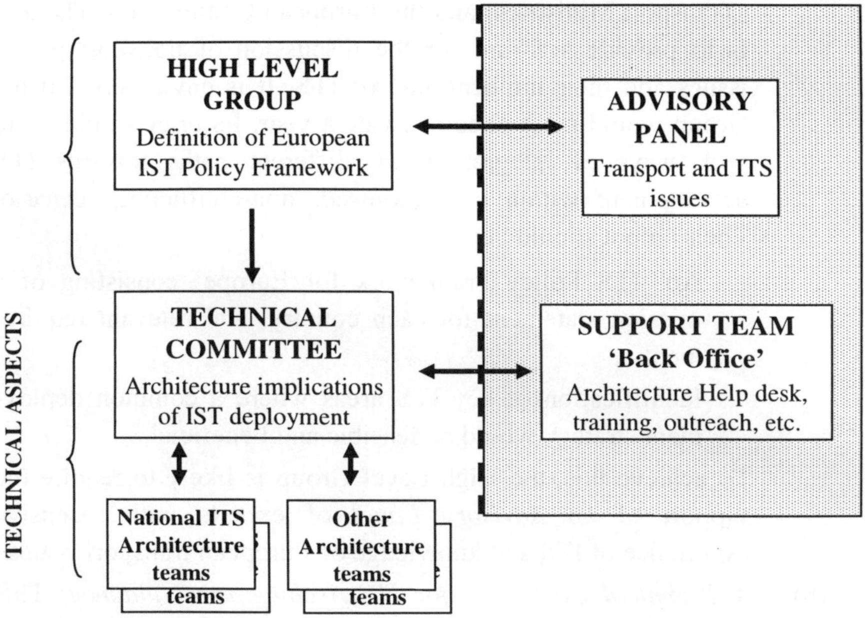

Figure 16: Proposed structure for ITS Policy Framework

7.1.6 *Recommendations*

The proposed strategy to foster an ITS Architecture Framework for Europe comprises the following three main principles:

1. Setting up of a mechanism to create an 'ITS Policy Framework';
2. Member States of the European Union should play a major role in future activities regarding ITS Architectures;
3. Technical support for architecture developers and users should be provided by a centrally-organised and funded body.

To make this possible, it is recommended that the European Commission should promote the institution of the following organisational elements:

(a) *A High Level Group for the harmonisation of ITS policy:* This would be made up of representatives of Member States

(Transport Ministries) and the European Commission. The aim is to provide a forum for the discussion of transport policy issues and their implications for ITS. It is envisaged that the Group would meet once or twice a year. Its brief would be to seek ways of integrating the different paths towards ITS development within a harmonised, non-conflicting scenario. Their output should be:

- An 'ITS Policy Framework for Europe' consisting of a mission statement for each country, and relevant requirements;
- Identification of key ITS areas where a common deployment strategy would be feasible and beneficial.

To achieve this, the High Level Group is likely to require the support of an *Advisory Panel* of experts with extensive experience of ITS and knowledge of European transport issues.

(b) *A Technical Committee for ITS Architecture Guidance:* This should consist of representatives of national and other relevant architecture teams. Its aim is to 'translate' the Policy Framework defined by the High Level Group into architecture requirements, and co-ordinate the necessary actions, including expansion and updates of the European ITS Framework Architecture and overseeing the incorporation of elements from national or other ITS architectures.

(c) *An Architecture Support Team or Technical 'Back Office'*: This team will have the responsibility of providing technical support to the national architecture teams and ensuring that new developments are compliant with the European ITS Framework Architecture and follow the established methodology. The role could also include the organisation of outreach activities and training workshops (i.e. similar to that of the FRAME projects).

One way of formalising the necessary commitment by European Member States to the organisational structure described above would be to draw up a Memorandum of Understanding. This would consist of a statement of willingness by the Member States to adopt a co-operative approach to the harmonisation of ITS deployment policies in Europe with the aim of promoting the interoperability of ITS applications, and to endorse the European ITS Framework Architecture as a basic tool for achieving this.

7.2 Radio-Navigation

This section examines the ability of the main space-based and terrestrial radio-navigation services to meet the location requirements of transport-related activities. It considers, in particular, the potential impact of the forthcoming GALILEO-based services and also looks at various augmentation techniques already available or expected in the near future. The main technical and policy issues that still need to be resolved are outlined, and some recommendations put forward to ensure that the needs of current and future ITS applications are satisfied.

To ensure that the format and content is consistent with work carried out elsewhere, the approach and parameters adopted here are based on those used for addressing radio-navigation requirements at the United Nations International Maritime Organization.

7.2.1 Background

Efforts have recently been made to raise the profile of radio-navigation requirements of terrestrial mobility to a similar if not equal footing to the air and maritime sectors. This has brought up the question of the vulnerability, security and equality of radio-navigation services and information services dependent on radio-navigation data.

In the past, most users of radio-navigation services came from within the professional maritime or aviation community. They were taught that the derived position information was to be treated with caution and that it was unwise to rely on a single radio-navigation service. In other words, they were supposed to treat all such services as ‘aids’ to navigation. The users of such services therefore had to be experienced and, above all, cautious due to shortcomings in the sensing equipment used and the poor reliability of some types of service.

It could be rightly claimed that we are now entering a new era for radio-navigation marked by radical changes not only in the level of service, but also the way in which the information is used and, consequently, the type of user. Modern receivers now incorporate software which greatly simplifies the human-machine interaction, automatically providing latitude and longitude, or even plotting a position directly on a map display. Radio-navigation services have become much more reliable, which means that users can depend on the information, rather than simply using it to support their own decision-making processes.

One of the results is that the number of users is set to explode from a few hundred thousands to billions. New communities in the transport field will include leisure and business activities associated with inland waterways, road travel, railways and emergency services, as well as the more traditional aviation and maritime areas. Millions of private individuals may well have radio-navigation capabilities incorporated in their mobile telephones and be able to utilise the information in many diverse applications, ranging from location of road vehicles or waterborne craft to navigation while walking or cycling. If radio-based location systems and other radio-navigation dependent information services such as these are available at reasonable cost and offer good reliability and accuracy, it is likely that many more applications not yet identified will emerge.

The numerous applications now becoming radio-navigation dependent have very different operational characteristics. While simple, static position information is sufficient for some, others need information on direction, dynamic velocity or vehicle/infrastructure status from one or many users within an operational envelope. In other cases the position information is only of use when used in conjunction with a topology engine that tracks the user within an envelope that might be an ‘allowable’ area or route. In terms of location accuracy, while a few services require very high precision (to within a few centimetres), for the majority of services medium levels of location accuracy are sufficient, but with high levels of reliability and availability over a very wide area.

7.2.2 *Vision*

We are living in an age in which technological and societal changes are creating a growing dependence on decision support systems that help to improve environmental protection, safety and efficiency. There is already a great appetite for tools that offer greater precision and integrity. In this context, radio-navigation services seem set to play an even greater socio-economic role as more and more critical transportation, environmental and safety processes become heavily dependent on them.

In the future, it is foreseen that a whole range of location services will be available through a single tool which gives support in all phases of a journey or mission, i.e. for navigation, positioning, tracking, as well as charging for use of the infrastructure or travel-related services. A person will have navigation assistance when driving into an unfamiliar city, in locating a facility such as a hotel, marina or airport, then continue the trip by a different mode (train, boat, plane), being supported throughout by location services which are not only reliable and accurate, but also presented in a similar fashion, creating a sense of safety, comfort and trust in the information provided.

Road and rail transport will be served by the same type of services used to give navigational support to vessels entering or leaving port: there will also be special functions such as goods tracking, and navigation for tourists or disabled people. Technology advances will enable receivers to 'upgrade' themselves, like present day PCs, with software which makes available new radio-navigation or telematics services as and when they become available.

Different levels of service will be provided for different ITS applications, depending upon whether they are considered 'safety critical' or 'mission critical'. The former can be defined as those that use services which have an impact on life or health; the latter are those with environmental or economic consequences.

This picture assumes that the fundamental policy decisions will have been made, and regulations and standards drawn up in order to ensure that applications are harmonised. Though requiring a large initial investment, with institutional support, it will be possible for cost recovery by service providers to be shared over many different applications.

7.2.3 State-of-the-art

Radio-navigation services consist of a number of components ranging from those that supply the location information to those enabling interactive decision support between users and the infrastructure serving them. They include:

- Stand-alone terrestrial and space-based radio-navigation services: GNSS (Global Navigation Satellite System – GALILEO & GPS), LORAN, Mobile Telephone Network solutions;
- Space-based and terrestrial services which enhance the accuracy, integrity and reliability of stand-alone radio-navigation services: EGNOS, Carrier Phase (Real-Time Kinematic), EUROFIX, IALA Differential GNSS beacons;
- Short Range 'affordable' local components for high accuracy, restricted range applications (Pseudolites);
- Systems for the measurement of vehicle and human motion, and map-matching products;
- Communication devices and services for promulgating and exchanging information from and between mobility users and the decision-support infrastructure;

Further necessary elements of successful location services are:

- Information protocols and architectures required for exchange of information from and between users and infrastructures;
- Policy instruments and standards to ensure European and, where necessary, worldwide interoperability of tools and products.

Radio-navigation services

Radio-navigation of the future will be made up of base-line services and/or technologies including:

Global Navigation Satellites System (GNSS) services:

- GPS and, in the future, GPS2 and 3
- GLONASS
- GALILEO

Terrestrial services:

- LORAN/Chayka (possibility of convergence)
- OTDOA (Observed time difference of Arrival) - Handset
- EOTD (Enhanced Observed Time Difference) - Handset

Indoor Services:

- Bluetooth
- WiFi

These base-line services can be augmented in various ways to improve their integrity and accuracy:

- GPS can be augmented by EGNOS (in Europe), WAAS (in the United States), and EUROFIX (LORAN-based WAAS in the United States) or by local RTK and medium-range differential stations (DGNSS) and space-based RTK;
- LORAN can be augmented by LORAN Hi Fix;
- Through the use of Pseudolites.

7.2.3.1 GNSS

Space-based positioning and navigation systems can provide meteorological, passive, three-dimensional position, velocity, and time data worldwide. In recent years the GNSS user community has expanded exponentially, with rapid growth occurring in all modes of transportation. This growth is expected to continue.

Formerly, the only GNSS service to be widely available was the American GPS system served by the NAVSTAR constellation. Though the Russian Global Navigation Satellite System (GLONASS) shares the same principles, its constellation consists of only 11 active satellites, which means it is virtually impossible to use as a stand-alone system. However, receivers able to decode both GPS and GLONASS signals have an accuracy advantage which is particularly useful in the urban environment, as they can exploit a larger number of satellites. Though dual receivers are currently available, they do not have a wide diffusion as their cost is very high (some thousand euros). This system is therefore not particularly attractive for telecom manufacturers.

As many transport applications are safety-critical, mission-critical or both, the radio-navigation service must be able to give authentication and integrity information. Stand-alone GPS is not able to do this, and although EGNOS can give integrity information, at present there are only a limited number of applications for which receivers capable of receiving EGNOS are available or practicable (due to size). GALILEO will be able to provide authenticity and integrity information as, eventually, will GPS 3.

Over the next few years Europe, with co-operation from China, will be commissioning its own GALILEO service which will operate along with GPS 2 (available from 2007) and GPS 3 (in 2015), as well as GLONASS. The first element of GNSS modernisation will enable civil GPS 2 users to correct for ionospheric errors using a second frequency in addition to the current signal. These corrections, when combined with switching off Selective Availability (SA), will enable equipment that meets benchmark standards to achieve horizontal accuracies in the 4-metre range. There will, in addition, be a third civil signal for safety-of-life applications.

The GALILEO service, planned to come into operation in 2008, will offer an independent, global, European-controlled satellite-based navigation system which will be complementary to GPS and GLONASS. It should provide position information with higher levels of accuracy than GPS 2, and will provide a number of services to users equipped with GALILEO receivers. It is proposed that GALILEO should offer differing levels of service to suit differing needs, including:

(a) Open Services (OS), free to all users, providing positioning navigation and timing performances comparable to or better than those of existing and planned global navigation systems;

(b) Commercial Services (CS), based on the open services signals, but providing value-added positioning, navigation and timing data to users (e.g. integrity), with a liability regime;

(c) A number of certified public interest services, such as Safety of Life (SOL) for transport and other safety-critical applications, Public Regulated Services (PRS) for the enforcement of Member States' and EU policies implementation, and Search and Rescue (SAR) services complementing COSPAS-SARSAT for the detection of distress alarms from user beacons. A further payload might facilitate two-way SMS communications.

These last categories are of particular interest for the mobility sector.

7.2.3.2 LORAN-C, CHAYKA and NELS

LORAN-C was originally developed to provide military users with a radio-navigation capability with much greater coverage and accuracy than its predecessor (LORAN-A). It was subsequently selected as the radio-navigation system for civil marine use in the US. Within north west Europe it is run as the NELS (North West European LORAN-C System) and is partially available within the Mediterranean under the SELS (Southern European LORAN-C System), which, however, lacks stations in Spain and Turkey. LORAN-C can also be used for precise time interval and highly accurate frequency applications.

The LORAN-C signal can be modulated to broadcast differential GPS correction data and GPS integrity information. In the US it is used as part of the LORAN WAAS integrity service; in Europe, as part of EUROFIX. However, the US and European services are not compatible. LORAN-C offers the advantage that the signal can be received even inside buildings. Over much of the northern hemisphere, LORAN-C or the similar Russian CHAYKA service provide coverage.

Some European Member State administrations are convinced that dependence on one type of service (GNSS) leads to unnecessary vulnerability and that LORAN would be a suitable candidate for a complementary service. At a relatively low cost, LORAN can be extended to give good accuracy coverage over all of Europe, as well as the surrounding seas and airspace.

7.2.3.3 Radio-navigation using mobile telephone network

Taking advantage of the positioning potential of cellular networks, a number of mobile phone positioning systems are being developed. Several commercial products are already on the market, mainly in the US, and especially in the Emergency Call Services field. The location methods used to locate a mobile telephone can be split into two categories: network-based solutions and handset-based solutions. The combination of both types (hybrid solution) is also used as it gains the advantages of the two techniques while limiting their drawbacks.

Bluetooth systems are being used for indoor applications within buildings, ships and trains for location of equipment, freight and people.

7.2.3.4 Inertial systems and dead reckoning

Many transport applications require a position or velocity update that is independent of radio-navigation services. There are occasions where the user is not within coverage of satellite or terrestrial services due to geographic location, in the shadow caused by tall buildings, mountains, canyons or tunnels, or during an interruption to a service. A variety of velocity sensors is on the market including vibrating gyros, Silicon single- and three-axis accelerometers, advanced accurate solid-state fibre optic true-north seeking gyro-compasses, 'solid-state' micro-machined quartz angular rate sensors, linear servo accelerometers and MEMS low-cost, high accuracy Silicon Micro-Ring Gyro inertial navigation systems (about 2 cubic centimetres in size). The prices range from a few tens of euros to hundreds of thousands of euros.

The quest today is to find ways of using the simple inertial measurement unit (IMU) with rough sensors for precise navigation. Companies worldwide are developing motion sensors, which are low cost inertial devices using cheap compact sensors. However, they have weak stand-alone accuracy and poor run-to-run stability and as such are not suitable as a sole system and require periodical updates.

7.2.3.5 'Real Time Kinematic' (RTK) terrestrial and space-based services

RTK is a high precision carrier phase-based positioning system using GPS dual frequency signals. The term Real Time Kinematic (RTK) is often used to describe carrier-based positioning systems that employ static reference station(s) and moving receiver(s).

Service infrastructures for carrier phase reference observations have been installed in several countries, but it is unclear how successful these are in real-time applications. Whilst excellent accuracy can be achieved in favourable environments, current GPS-based RTK systems have performance limitations in relation to the baseline length (reference to user separation), the need for high bandwidth data links, and their overall availability and robustness. There are several applications that would potentially benefit from a robust service providing centimetre accuracy. The future introduction of satellite transmissions on three frequencies, and also the future GALILEO constellation, will give processing advantages offering better performance and reliability, probably delivering accuracy to under 20 cm at a frequency of every few seconds.

For most mobility applications, terrestrial services are probably better due to their greater vertical accuracy. Currently within Europe the project EUPOS is investigating the expansion and harmonisation of services throughout Europe. EUPOS will offer two levels of service: one giving accuracy of a couple of metres, the other to one or two decimetres (horizontally and vertically). Such services already give accuracies of a few centimetres in good coverage areas.

7.2.3.6 Pseudolites and synchrolites

An increasing number of applications require precise relative position and clock offset information. In situations with limited or no visibility of the GNSS satellites or RTK constellations, ground transmitters that emulate the signal structure of the GPS satellites (pseudolites) can be used as additional or replacement signal sources. Transceivers (which transmit and receive GPS signals) can be used to improve standard pseudolite positioning systems. If their locations are known, transceivers can be used to remove the need for the reference antennae typically necessary in standard differential systems. In addition, transceivers mounted on vehicles can allow continuous inter-vehicle positioning without the presence of signals from GPS satellites.

Pseudolites are ideal for covering areas that are in shadow from the RTK or GNSS service, or where interference from re-radiation from structures or other sources makes reception of the service impossible. They may also be used where centimetre-level position information is required for horizontal or vertical accuracy, but there is not enough coverage demand to warrant a Carrier-phase Differential GPS (CDGPS) (Real Time Kinematic) installation.

The following table shows the positioning accuracies offered by the different approaches described above.

Table 7: Approximate positioning accuracy of existing and emerging services

Service/ technology	Accuracy (m)	Indoor	Tunnels & Canyon	Outdoor
Cell ID	100-35,000	Restricted	Restricted	Yes
TA	550	Restricted	Restricted	Yes
TOA	125	Restricted	Restricted	Yes
AOA	125	Restricted	Restricted	Yes
EOTD	50-150	Restricted	Restricted	Yes
Bluetooth	<30	Yes	(Portals only)	No
GNSS	20 (4 in future)	GALILEO & AGPS	No	Yes
DGNSS	1-7	GALILEO & AGPS	No	Yes
LORAN	<500	Yes	No	Yes
Pseudolites	0.01	Yes	Yes	Yes
RTK	0.03-0.2	No	No	Yes

7.2.4 Issues

7.2.4.1 Precision and performance

Current non-augmented satellite-based radio-navigation services, such as GPS, do not meet all requirements for many applications. A user must have at least five satellites in view above a mask angle of 7.5 degrees in order to provide Receiver Autonomous Integrity Monitoring (RAIM). This condition is not always satisfied with the existing GPS constellation, resulting in so-called 'RAIM holes' and limiting GPS to use as a supplemental navigation system. To meet the requirements for Fault Detection and Exclusion (FDE), at least six satellites with good geometry are necessary. However, the adverse effects of these variances may be substantially reduced, if not eliminated, by differential techniques. In such operations, a reference station is located at a fixed point (or points) within an area of interest. GPS signals are observed in real time and compared with signals expected to be observed at the fixed point. Differences between observed signals and predicted signals are transmitted to users as differential corrections to upgrade the precision and performance of the user's receiver.

EGNOS, WAAS, and MSAS are safety-critical systems designed to augment the existing satellite services. The United States WAAS service provides GPS augmentation of much of North America and the Hawaiian Islands. Japan is developing the Multi-functional Transport Satellite (MTSAT) for the Satellite Based Augmentation System (MSAS) to provide GPS augmentation of some of the Far Eastern area. The European EGNOS service will initially provide augmentation over the European area, and there are plans to expand coverage.

All of these systems will be interoperable, with the result that the user will require a single receiver to use all three. Within the areas mentioned, navigation performance is expected to be similar to that in the EGNOS Core Coverage Area, i.e. around 5-7 m accuracy. It is important to note that whereas EGNOS was specifically designed to provide a multimodal service, WAAS and MSAS were developed as aviation systems. The

WAAS is provided by a signal-in-space and via the US LORAN service to airborne and terrestrial WAAS users to support precision navigation.

7.2.4.2 Integrity

There is one caveat when considering SBAS (Satellite-Based Augmentation Services). Though they give an increase in accuracy or integrity of information, they can only provide it for the GPS satellites within their own horizon. This may not be significant for terrestrial users in Europe, but when journeying east for example, there may be occasions when some of the satellites within the user's horizon (and being used for a position fix by the user's receiver) are not covered by EGNOS GPS monitoring. The information provided will therefore only give partial integrity, resulting in a mix of reliabilities for the computed fix. This is, of course, critical for applications that are dependent on integrity (road charging and vehicle insurance, for instance). When GALILEO is available, this problem will disappear, as all GALILEO satellites will provide authentication and integrity information. GPS 3 will also provide integrity information, but this constellation is not due until 2015.

Both the EUROFIX system in use within Europe today and the WAAS LORAN service, shortly to be implemented in the US, require the LORAN infrastructure. Despite discussions on providing EUROFIX signals by LORAN-C stations in Europe and the United States, and by CHAYKA stations in the CIS (Commonwealth of Independent States), the US have decided on a system developed by Stanford University that is not compatible with EUROFIX.

The EUROFIX system has an uncertain future in Europe, since it is dependent on the LORAN infrastructure, which requires political direction (i.e. a decision on funding to keep the NELS LORAN service commissioned). To ensure maximum practicable coverage, the co-operation of Japan, South Korea, China, Saudi Arabia and India will be necessary.

There are many differential GPS (dGPS) marine stations in Europe located mainly in the Baltic, North, Barents and Irish Seas as well as Icelandic waters and along the Atlantic coastlines. When the network is complete, stations will provide overlapping coverage in most areas, and will cover the Canary Islands and the Western and Central parts of the Mediterranean Sea. To provide coverage at the Eastern end of the Mediterranean Sea would probably require an additional ten DGNSS stations.

In addition to marine coverage, work is ongoing to investigate the quality of radio-beacon DGNSS service over European land masses, particularly in the UK and France. However, it should be noted that a significant increase in the number of inland beacons will be needed for large countries with small coastlines (e.g. Germany).

Carrier-phase Differential GPS, CDGPS (Real Time Kinematic), can provide centimetre level position information. It is becoming increasingly important for autonomous vehicle control.

7.2.4.3 *'Vulnerability' (dependency on one gender of service)*

Many believe that critical application (safety, environment and economic) should not rely on one gender of radio-navigation service alone. Because GALILEO and GPS share very similar frequency spectra (i.e. of the same gender), both could be vulnerable to spoofing or, much more likely, jamming. This could occur as an act of aggression against a sovereign member or as part of terrorist action. It is also evident that should a threat occur against the national security of any individual EU or NATO Member State, there could be a demand for space-based radio-navigation services, such as GPS or GALILEO, to be made unavailable. There is also a remote possibility that NATO members who do not have sovereign control over GALILEO might take their own steps to remove the usability of the signals in regions of military action for short-term tactical or medium-term strategic purposes. It is also possible for technical problems to lead to interruptions to the space-based signal, as with the GPS WAAS service in 2002, or to provide erroneous information, as with GPS in January 2004.

For all these reasons, it is important to have terrestrial services that will continue to serve the community in the absence of useable space-based signals. This means that a vehicle or vessel should be capable of receiving information from more than one type of service.

LORAN-C is available widely in the northern hemisphere, but is at present blighted by poor political direction. Originally the service was planned to be shut down in 2004. However there has been a movement that has ensured it continues operating for the time being. For LORAN to be really useful for terrestrial coverage it needs to be extended to complete the SELS (Southern European LORAN Service). It is also true that LORAN-C is used in other parts of the world (including USA). Because it is compatible with CHAYKA, there is a possibility of complete coverage of continental Europe. It is also the only terrestrial service that has good coverage over large sea areas and remote, scarcely-populated areas.

Subsequent to the VOLPE report and the 9/11 atrocities, LORAN is being considered by the US as a relatively low cost solution to provide a non-space-based complementary service. Studies have been conducted that show it could be cost effective to use LORAN providing that a pan-European initiative could be found to fund the cost of commissioning new stations required as well as continue operation and maintenance of the infrastructure as a whole. In view of the possibilities of combining LORAN with CHAYKA there is need to determine whether the service is viable as a back-up and would provide a useful service for all transport and mobility users.

Manufacturers are unlikely to develop and make low cost receivers commonly available unless Europe has a clear trans-European policy direction that LORAN would continue for the foreseeable future, and SELS (Southern European LORAN Service) would also have to be fully implemented within the Mediterranean Sea.

7.2.4.4 Map matching

There is a difference between the WGS84 datum, which is the mathematical model of the world's surface used for GNSS, and that of many maps and charts. In most cases, the differences are known to be less than 200 metres, although differences as great as several miles have sometimes been reported. Cartographic offices are attempting to refer as many new maps and charts as possible to WGS84, but there remain many parts of the world where the necessary information does not exist and where positions obtained from satellite navigation receivers cannot be used without adjustment.

In most European areas the difference is known. However, the shift correction provided is an average for the area of coverage and therefore may not be consistent with the shift correction of another map covering part of the same area. This may result in the position plotted being slightly different. Older maps and charts that have not been shifted to WGS84 will need a mathematical shift correction to be applied before plotting.

Though many manufacturers of GPS receivers are now incorporating datum transformations into their software, there are unfortunately many cases where a single transformation will not be accurate for a large regional datum. For example, the relationship between the WGS84 datum and the European datum (1950) is very different in the north and the south of Europe. For many GPS applications this may not be problematic, but it is an additional source of error and may be significant for a safety critical application for which, if differential GPS is being used, a position accuracy of just a few metres is required.

It cannot be assumed that all maps and charts in a region refer to the same datum. Although most metric charts of the European mainland and waters are referred to the European datum (1950), many charts also use a local datum. In addition, as there are no international standards defining the conversion parameters between different horizontal data, those used by the current and future GNSS devices may be different.

Current positioning with dGPS is usually more accurate than position-fixing used for surveys conducted before 1980. The result is that although users may know their own position to an accuracy better than 10 metres, the positions of objects in the vicinity may only be known to an accuracy of 20 metres or worse. Unfortunately, it will be many years before all areas are re-surveyed and all maps and charts revised.

A satellite navigation receiver may give a position to a precision of three decimal places of a minute, but that does not mean that all its positions are accurate to 2 metres or that the resulting position is compatible with the positions of objects shown on modern maps or charts (paper or digital) which may have been produced 100 years ago or more.

Extreme caution therefore has to be exercised when combining satellite-derived position information with paper maps and charts, or even GIS tools. It is important that the user is aware of potential differences and compensates accordingly.

ITS applications are still evolving, but will in the future provide remote decision-support and integrated vehicle/traffic management, and one day even automated vehicle guidance. This means that both the driver and infrastructure operator will have to rely on position information from radio-navigation systems. Unless the datum issue is tackled this information could be inaccurate. Such discrepancies could create serious safety problems, especially on congested roads. Future Advanced Driver Assistance Systems (ADAS) applications, for example, will require 50 cm accuracy and therefore a map accuracy of 25 cm. Presently, most road maps give about 5 metre accuracy. This problem is not restricted only to paper maps, but is further frustrated by the possibility that map and chart projection discrepancies might occur between horizontal and vertical data of differing 2- and 3-dimensional GIS electronic chart and map standards.

7.2.4.5 High levels of accuracy

Some applications are reliant on sub-centimetre accuracy. In general, these applications do not require a latitude and longitude position, but reference from one point to another. This is typical of ADAS applications, for example. However, every situation where such accuracy is required is different. In some cases, if there is coverage by two or more RTK stations, the required accuracy might be achieved. In other cases, the use of pseudolites may be required.

Pseudolite technology can provide the necessary accuracy, but there is a very real possibility that the transmissions required might also interfere with normal non-pseudolite enabled receivers. Therefore, before implementing this technology, steps must be taken to ensure that their signals will not spoof or jam normal GNSS receivers. In the future, the arrival of GALILEO and GPS 3 will mean that all users will require new receivers. It is, therefore, important that new receiver equipment used for navigation or reference applications should be able to use RTK and pseudolite services that are interoperable throughout the world.

7.2.5 *Future opportunities*

The entry into service of the GALILEO constellation planned for 2008 will permit many of the shortcomings of existing radio navigation systems to be overcome. While many benefits will derive from the features of the GALILEO system itself, others will result from its use in combination with GPS and GLONASS. Among the expected advantages are:

- *Higher accuracy and update frequency*, due to the addition of the GALILEO satellites which can be used in conjunction with GLONASS and GPS;
- *Suppression of multipath interference*. Again the larger number of satellites can reduce the impact of interference caused by re-radiation or reflection of GPS signals which occurs in the vicinity of large steel structures or communication antennae. The use of E5A and E5B signals together as a super-wideband signal can produce an improvement in reception. Since cities and ports are particularly vulnerable, this is very relevant for ITS applications. Local augmentation services may still nevertheless be required, especially for high end/marginal situations;
- *Correction of ionospheric interference* is possible with the use of EGNOS, which permits mapping of the ionospheric path of each satellite's signal and can therefore calculate the difference between pseudo range and actual range, enabling a map to be produced of the electron content of the atmosphere. This information can be applied as a correction for ionospheric

error. As GPS 2 and GALILEO will be providing more than one signal, a receiver will be able to provide its own corrections;

- *Authenticity and integrity information* from GPS satellites can be improved by the addition of EGNOS, but this applies only to those which can be monitored (i.e. not when they are outside or only partially within EGNOS coverage). Through the addition of three more signals in space, EGNOS will improve the predictability, availability, accuracy and continuity of information, which GALILEO will improve yet further. With more satellites, the accuracy and reliability of the calculated position can be greatly enhanced. The information will therefore be more trustworthy with EGNOS, and even more so with GALILEO;
- *World-wide coverage* will be possible with GALILEO, whereas it is not provided by EGNOS and other SBAS (Space-based augmentation services). GPS 3, like GALILEO, should provide its own integrity data, but until GALILEO services are available, GPS 3 will not be able to supply integrity and ionospheric data when it is out of coverage of MSAS, WAAS or EGNOS;
- *Political independence.* Until GALILEO is in operation, Europe will have no politically independent services, since GPS or GLONASS are the only services available at present, and EGNOS is dependent on them.

Despite these many potential benefits, it is necessary that steps should be taken to provide back-up services which would ensure coverage in the case of technical problems, jamming or spoofing.

Back-up services

The requirements for back-up terrestrial services (suitable for medium- and short-range high-accuracy applications as well as longer range less accurate services) are met by LORAN-C. This is currently the only medium/long range service widely available in the northern hemisphere, but is at present blighted by poor political direction.

LORAN has recently enjoyed an extension of life beyond 2004. France in particular has been very active in supporting LORAN and in providing the possibility for some extension of coverage. Other Member States have a non-biased policy on LORAN, whereas others are positively seeking a European policy decision to find an instrument for the continuation of NELS (North European LORAN Service) at least until a viable alternative emerges and, if possible, the reinstatement of two stations to extend coverage into the Mediterranean to complete the SELS (Southern European LORAN Service).

Though GPRS (mobile telephone network) techniques are emerging in conjunction with the e-112 service, and could be used where coverage exists, they are not yet widespread enough and on their own do not offer longevity of high accuracy (high accuracy use of these techniques requires the combined use of GNSS).

7.2.6 *Recommendations*

If the opportunities outlined above are to become a reality, it will be necessary to take some strategic policy decisions at the European level to establish regulations and draw up standards which will help achieve harmonisation. Though initially this will involve a huge investment, and is likely to require private funding, cost recovery by service providers could be distributed over many different applications and through various institutional instruments.

There is an urgent need to define a policy for Europe that will provide the instruments to serve the needs of the mobility sector in the near and medium term. At the time of writing, the process of setting out a strategic plan for radio-navigation in Europe appears to have stalled, although a recently completed technical study provides the necessary basis from which such a policy could be derived.

A step which would undoubtedly favour the development of a coherent long-term plan would be the creation of a European Radio-Navigation Agency. At present, an agency is being set up to steer the development of GALILEO-supported services. We would suggest that serious consideration is given to the institution of a European body responsible for furthering the coherent development of all forms of Radio-Navigation. Such a body could include GALILEO-based services as one category of navigation service. The risk otherwise is to have to wait until the year 2011, when GALILEO is fully operational, before it is possible to negotiate other radio-navigation initiatives.

We outline below other priorities for future action:

(a) *Finding a terrestrial alternative to GNSS services, and examining the feasibility of LORAN and EUROFIX for transport applications.*

In light of the complete reliance on GNSS for radio-navigation coverage of all European land and sea areas, and the potential vulnerability of reliance on one type of service, Member States should be urged to lobby for examination of a terrestrial alternative to GNSS. For this reason, urgent action is required to establish the viability of LORAN and EUROFIX for all transport and mobility communities. However, if pseudolites are to be used, although this will benefit applications not covered by space-based services, there is a very real possibility that the transmissions might interfere with normal non-pseudolite-enabled receivers. Therefore before implementing pseudolite technology, steps must be taken to ensure that the signals will not spoof or jam normal GNSS receivers.

(b) *Promote the Europe-wide implementation of DGNSS services complemented with back-up from RTK.*

European Member States should also be urged to support and, where necessary, provide funding and instruments for the harmonisation and implementation of Europe-wide DGNSS and RTK services and user receivers. Such a programme is needed to ensure total coverage of DGNSS and the availability of RTK where necessary. Though in most cases the accuracy from DGNSS will be sufficient, in critical congested infrastructures and areas of restricted accessibility higher accuracy solutions will be required. It should be borne in mind that many applications needing centimetre accuracy only require knowledge of the difference of velocity relative to an infrastructure, and not an absolute position.

(c) *Promote interoperability of receivers and ensure that critical services are protected by authenticity procedures.*

Users' receivers must be capable of transition between services and, when interfaced to automatic identification devices required for security or traffic management or control applications, be able to deliver information of a high quality to others within the system.

It will be necessary to ensure that any user who needs to interface with critical infrastructures or applications must have services with authenticity information to ensure that the signal used is a bona fide GNSS signal and not spoofed.

(d) *Find ways of ensuring security and protecting signals from hostile interference, including setting up of back-up services.*

Mechanisms should also be put in place to permit the removal of radio-navigation signals which are malfunctioning or affected by hostile action, in order to ensure that the user remains safe and the infrastructure is not susceptible to major incidents. Vulnerability studies have been carried out for GNSS and many have suggested LORAN as a back-up service. However, it will be necessary to conduct a similar analysis of LORAN as a terrestrial-based 'physical' infrastructure is more easily put out of action than a space-based infrastructure.

In deciding priorities for action and the level of service quality, it will be important to determine which processes and services are 'safety critical' and which are only 'mission critical'. The former can be defined as those that use services which have an impact on life or health, whereas the latter are those that have environmental or economic consequences.

7.3 Education and Training

Enormous changes have affected the world of transport over the last few decades. There now exists an abundance of new technologies and techniques which are applicable to transport networks and operations. However, as previous chapters have indicated, the existence of the technology does not by itself ensure the successful implementation of ITS services. Among the essential supporting factors is the availability of the skills, knowledge and experience required for the design, development, deployment and maintenance of such systems.

The lack of investment in human resources, and the 'newness' of many of the skills, has resulted in a shortage in much of Europe of professional people with expertise in the many facets of ITS. Not only is there a lack of technical knowledge, but also a low level of general awareness of the costs and benefits of ITS. Given that many major policy issues – from road use charging to intermodal transport and questions of safety and security – have ITS-related implications, this is a serious shortcoming.

There exists, nevertheless, a great deal of valid experience in Europe in the form of personal knowledge and research results. These are often lost, or not fully exploited, due to the lack of a systematic mechanism for their dissemination. Despite the growing number of ITS-related courses in universities, and training seminars for mid-career professionals, there is currently no overall co-ordination at European level, and few attempts to share teaching resources or exchange experience.

7.3.1 Vision

In our vision of an ideal future, students interested in taking up an ITS-related career, researchers and also professionals wishing to refresh and extend their ITS knowledge, will have easy access to a range of high quality training courses and supporting facilities, including virtual laboratories, information on 'best practice' and the possibility of field visits to successful ITS schemes.

The basic courses will give a thorough grounding in the skills necessary for future ITS professionals, covering policy definition and project management as well as technical subjects. Other more specialised courses will permit a focus on particular areas of knowledge.

To support them in their choice, comprehensive information about the opportunities for ITS training and education in Europe will be readily available from a dedicated website. This centralised source will supply details of curricula in institutions offering full-time courses, as well as news of short courses and events of interest. On submitting a personal profile and requirements, an online support service will give suggestions on the choice of appropriate options.

University students will also be able to pick up brochures presenting the various specialisations (MSc and PhD) leading to careers in ITS. Those interested in following ITS courses, but living in regions where these are not available, will have the option of following modules in other universities, enrolling on web-based courses, or using forms of e-learning, all offering credits towards their degrees.

All of these supporting services will be developed and made available under the auspices of a European ITS Training and Education Network. Lecturers who work in institutions belonging to the Network will have the benefit of support in the preparation of courses by having access to modules on ITS-related subjects. These will include material on all the major ITS areas. There will also be collections of real-world case studies and of recent research results.

Such material will be available either on payment or through an exchange arrangement based on credits given for contributions to the 'pool'. Since all material will have to be approved by a Technical Committee, it will carry a guarantee of scientific quality. The Network will also offer guidelines on ITS curricula and the opportunity to exchange ideas.

While the leading members of the Network will be educational and research institutes, associate membership will be open to ITS bodies, public administrations and industry. These will be invited to contribute suitable modules on ITS applications and products for use in training programmes, and also to participate in seminars on ITS topics.

A further function of the Network will be the delivery of short courses (one or two day workshops) on key ITS issues or subjects of international relevance, such as the European ITS Framework Architecture. These courses will be prepared either by recognised experts or Network members and will be available at a moderate cost to professional staff working for local authorities, transport operators, and firms in the sector.

The intention is to offer in-depth investigation of the subjects, aiming to give an academic treatment which is as objective as possible, rather than a political or industry-oriented approach. As with the training material, courses will be officially recognised with guaranteed quality standards and certification.

Senior executives working for public administrations or local authorities in positions carrying responsibility for transport policy decisions, will be able to take half-day 'awareness' seminars designed to explain the costs and benefits of ITS. They will also be offered guidance in the training facilities for staff involved in ITS-related projects, and encouraged to give funding to personnel attending certified courses. At the same time, authorities which have themselves already implemented successful ITS initiatives will be encouraged to make them available for site visits by students and trainees.

7.3.2 State-of-the-art

7.3.2.1 Universities and institutes of higher education

The availability of university-level education in ITS is not uniformly spread across Europe. While in some countries there are a number of institutions offering courses in ITS-related subjects, others have none at all. But even in the former case, very few universities offer a comprehensive curriculum specifically focusing on ITS.

Most courses are part of transport or engineering programmes (most often taught within civil engineering departments), and tend to touch only on certain aspects of ITS. The majority cater for Masters or PhD students with first degrees, typically in civil or electronic engineering, and sometimes information technology. The most usual pattern is for 20-40 hours of lessons to be included within a specialisation in traffic engineering or transport management.

Many new initiatives are nevertheless appearing. In Germany, for example, new chairs in ITS-related subjects have recently been created. ITS courses have also been launched in research institutions specialising in traffic engineering and traffic research. There is growing co-operation between universities, at least at national level, with joint arrangements being made to offer part-time modular courses leading to an MSc in ITS.

There are also exchange schemes operating at European level, although such arrangements generally involve a limited number of countries and have a limited duration. Some consortia of universities are now offering an 'international' MSc in European Traffic and Transportation (including some ITS elements) after three years part-time study, mainly via distance learning. Several universities also run courses for working professionals, and in a few European countries it is now possible to take a Masters in Transport not only through full- or part-time study, but also 'day release' or distance learning.

7.3.2.2 Training for professionals

A survey undertaken in the late 1990s in the United Kingdom indicated that only 12 % of the respondents (all members of the national ITS Association) had received formal training. The majority had picked up their knowledge in the working context or through attendance at seminars and congresses. One of the consequences of this (probably typical) situation is that the knowledge risks being partial and unstructured, with the lack of an overall perspective.

Employer-led initiatives sometimes take the form of in-house training, but more often arrangements are made for employees to attend courses organised by professional bodies or universities. Employers occasionally sponsor certificated courses, especially for younger personnel, in a local university, but the preference is usually for short (1-3 days) or very short (half day) courses on specific subjects.

Several European countries have institutions which focus specifically on running training courses for professionals in the transport field (examples are AFT-IFTIM in France and PTRC in the UK). In general, however, the ITS-related courses being offered by such institutions deal with specific areas, such as freight transport operations (ITS applications for logistics and fleet management) and vehicle applications (e.g. on-board systems for driver assistance).

It is increasingly common to find one or two day seminars being offered by the private sector on 'hot' ITS topics. Recent years have seen a spate of seminars on subjects such as road user charging and infomobility. These are often very expensive and in many cases, though informative, are prone to political, technical or commercial bias.

An area of demand not generally catered for in a satisfactory way is the training of the staff of public administrations (transport ministries, local and regional authorities), despite the fact that a good understanding of the costs and benefits of ITS is essential for informed policy decisions. Technical staff also need to be able to prepare tenders and commission ITS projects. Another evident 'gap' is the lack of refresher courses for mid-career professionals, who need to keep pace with new developments in the field.

7.3.2.3 ITS training initiatives at European level

A number of 'European' training initiatives have been carried out in recent years. Some of these are described briefly below. The list is by no means exhaustive, but is intended to provide some representative examples. While all involve co-operative efforts to deliver courses and/or create teaching resources in an international context, they illustrate some very different approaches.

In the early 1990s, a series of crash courses in ITS was organised in France by the ATRACC initiative as part of the EC's COMETT programme. One of the principle aims was to give wider dissemination to the outcome of EU research projects. In the view of participants, one of the most valuable features of these courses was the overview gained, in a short time, of topical aspects of ITS.

Each 3-day ATRACC course focused on a specific ITS area (e.g. Urban Traffic Control, Route Guidance Systems, Demand Management, Impact Assessment and Evaluation). Lectures were given in English by experts, and a 'Case Study' presented by someone with practical field experience. They attracted participants from different countries and types of institution (private firms, local authorities, ministries, etc.). However, unless external funding is available, the cost of participating in this kind of course can be high (in this case around €1000 each).

Training courses have also been organised as part of European funded research programmes in specific subject areas. One successful example was the training programme run by the FRAME-S project in the period 2002-2004. This consisted of a series of seminars and workshops which had the aim of raising awareness and understanding of the European ITS Framework Architecture. The sessions were free of charge and could be requested by national or other groups involved in ITS. They consisted of one-day seminars for ministry or local authority staff who needed sufficient knowledge of ITS architectures to support policy decisions, and two-day workshops for technicians needing to know how to develop ITS architectures themselves.

The aim of the EC project, PORTAL, which ran from 2000-2003, was to accelerate the take-up of EU research results on urban and regional transport. EC research project reports were used as a source of new teaching material, which was to be translated into the 16 languages of the European Union.

Although little of the content related to ITS, the experience was relevant in that it revealed a number of practical problems associated with using project results as a basis. It was found to be i) difficult to provide comprehensive coverage of a topic, ii) sometimes hard to establish ownership of intellectual property rights, iii) necessary to set up a quality assurance mechanism and to find ways of updating the material. A further problem was caused by the translations, which proved to be costly and, when carried out by non-technical people, not always accurate.

While material produced in this way cannot constitute a complete course, it can still serve as a very useful 'pool' of information to be customised by lecturers, or used for reference by students. The content most appreciated was reported to be the 'real-life' case studies and examples of best practice. The PORTAL material was made available on the Internet.

Short seminars on topical transport issues have also been organised by European projects, such as the 'Training Programme for Urban Professionals' funded by DGTREN. While these can be offered free of charge or at very low cost, they are inevitably limited in number.

In the research field, there are currently several European initiatives which include training activities. The HUMANIST project, for instance, is a Network of Excellence (2004-2008) whose objective is to create a virtual research centre which will examine all aspects of human-related design in connection with IT technologies applied to road transport. It is supported by ECTRI (European Conference of Transport Research Institutes), a body which actively promotes co-operation in surface transport research.

A general problem raised by EC funded courses is how to ensure continuation after project closure, and in particular how to make provision for updating the material produced. The preparation of future trainers can help but, in the long term, the continued relevance of content can only be ensured by creation of a permanent self-supporting mechanism.

7.3.2.4 Initiatives outside Europe

In view of their relevance and potential applicability to Europe, two initiatives regarding ITS training and education outside Europe are briefly described. These regard the academic network set up in 2002 in the Asia-Pacific region, and the Professional Capacity Building (PCB) programme in the United States. In both parts of the world, as in Europe, the lack of experienced teachers and of suitable training materials for ITS has been recognised as a problem.

Asia-Pacific Academic Network

As part of a search for new ideas and methods, ITS experts from Keio University in Japan decided to undertake a world-wide survey of ITS training. Following on from this initiative, a Symposium on ITS Training and Education was held in June 2002 in order to exchange information and discuss training strategies. This led to the launch of a 'Network of Excellence' for the Asia-Pacific region (including Japan, Malaysia, South Korea, Chinese Taipei, Peoples Republic of China, Thailand and Australia) with the aim of promoting and co-ordinating training, and disseminating examples of best practice.

One of the conclusions of the Japanese survey was that the introduction of ITS training requires a 'multi-step, multi-element' programme, which can bring together elements currently within different disciplines, and which can cater for differing needs. The latter include:

- Permanent courses for undergraduates and postgraduates;
- Training for ITS educators;
- Follow-up training for professionals in the transport industry;
- Continuing education for decision-makers in industry, as well as in local and central government;
- Specific vocational training for operators in the transport industry.

The analysis pointed out the international nature of the ITS field, and the consequent benefits of adopting a global perspective in teaching in order to promote harmonisation of technologies, specifications and standards. It was stressed that courses should consider not only ITS technologies, but also elements relating to project management and finance, such as:

- Cost benefit evaluations;
- Funding of ITS schemes;
- Questions of standards and international procurement;
- Quality assurance systems;
- Business models;
- Obligations of private and public sectors;
- User expectations and market analysis.

As well as these general principles, the concept of a 'macro-regional network' for training and education is clearly relevant to Europe.

The Professional Capacity Building Program

One of the most comprehensive initiatives for ITS training and education is undoubtedly the Professional Capacity Building (PCB) programme in the United States. The PCB is funded by the US Department of Transport and has an active partnership with (among others) ITS America, the National Highway Institute, National Training Institute, Institute for Transportation Engineers, and the Consortium for ITS Training and Education (CITE), which organises web-based training courses.

The PCB's aim is to provide information for anyone interested in improving their knowledge of ITS, at various levels of specialisation: students and professionals working in transport and other related fields, including those with non-technical as well as technical functions. Among the services offered by the PCB is a website which acts as a centralised 'one-stop' source of information on a wide range of ITS-related topics, among them:

- Catalogues with course descriptions (university and short training courses);
- Access to web-based courses (e.g. those organised by CITE);
- ITS Curriculum Guide (practical online support for constructing a personalised training programme);
- Quarterly newsletter and calendar of ITS events;
- Links to other related organisations and resources.

The classroom courses included in the programme are divided into three levels: i) 'awareness learning' (a general introduction to ITS), ii) 'core learning' (broad conceptual level of knowledge for all transport modes), iii) 'specialised learning' (deeper knowledge for specific deployments).

CITE was set up in the United States in the late 1990s, and now comprises around 80 universities and industry associations. Its aim is to provide a broad range of ITS training through distance learning. It has built up a series of interactive courses, delivered via the Internet, and is geared both to students and mid-career professionals.

By 2004, around two dozen courses were already available, including one full-semester course consisting of 11 modules on the fundamentals of ITS and Traffic Management, at a cost of $950. Other courses cost on average $275, and the two certificate courses around $600. A number of textbooks are available as support for the online courses. On completion of recognised courses, students are awarded credits, referred to as CEUs (Continuing Education Units).

One of the basic principles of CITE is that it should be self-supporting. Costs are kept as low as possible through co-operative arrangements and contributions, e.g. to course development, from consortium members. Revenue derives mainly from sponsorship, course fees and book sales. The organisational principles and relationship between the members are set out in a Business Model.

CITE offers a good example of the potential of distance learning for ITS. The US-oriented nature of the course content means that the material itself is of limited validity in the European context, but there are many features of the approach which could be applied in a European setting, as well as the business model and approach to course development.

7.3.3 *Issues*

The review carried out by ROSETTA brought to light a number of issues and problems relating to ITS education and training. These are examined below, and an attempt made to identify the opportunities which exist and ways of moving towards the vision outlined at the beginning of this section.

7.3.3.1 The scope of ITS

One of the difficulties encountered when attempting to develop a coherent and comprehensive teaching programme for ITS, is the lack of a clear definition of the scope of ITS. Being a relatively new area of study, ITS-related courses, especially in the university context, are at present embedded within different departments and faculties.

Although courses are commonly found in Civil Engineering departments, this is not necessarily the best 'home' for ITS. It could equally well be electronics, systems engineering, logistics, or other areas, depending on the intended focus.

Dealing with ITS in the working context clearly requires a range of skills and areas of knowledge, including technological aspects, but also management, legal, social and institutional issues. The definition of suitable curricula consequently calls for a readiness to break through institutional barriers and to adopt a co-operative approach.

One useful contribution at European level would be the preparation of guidelines for training programmes and university curricula, based on a careful analysis of the key competencies required. In the PCB programme, for instance, the required competencies have been divided into 'bundles' including systems engineering, institutional issues, technology and data management, planning and evaluation, and others (e.g. management of contractors, finance or communication skills).

Whatever the skills considered most appropriate for the world of transport today, it will be necessary to look critically at existing courses with a view to gradually remodelling curricula. It will, in the meanwhile, be essential to encourage as much flexibility as possible, by offering courses via a range of different channels and the 'mix and match' of teaching resources. Various ways of achieving this are discussed below. They include the creation of a pool of teaching modules at European level, the establishment of an international credit system, the development of web-based courses, and organisation of short crash courses.

7.3.3.2 Assessing demand

Before investing time and financial resources in the production of material and facilities, it would be helpful to have some knowledge of the market, i.e. the requirements of the various target audiences, appropriate methods of delivery for different types of course, supporting materials needed, etc. This is not easy to evaluate, especially as the demand for skills changes over time, with the evolution in transport policies, the introduction of new tools and technologies, and the general shift in requirements as the market matures, i.e. from the system design stage to implementation, maintenance and extension.

With regard to course content, several demand surveys have been carried out recently in Europe, but most regard single countries, or were not ITS-specific (i.e. referred to transport as a whole). They do, however, provide some very general indications of the type of training most requested.

High on the list of necessary skills identified by a recent UK survey were *system design and project design* (requiring a high level of technical knowledge); *project management and scheme assessment* (needing an overall view of ITS plus analytical business skills); *software engineering and system testing* (again technical know how). One of the major shortcomings was claimed to be lack of appreciation of the 'big picture' and of skills in system integration. Both are aspects which current trends in ITS deployment are making increasingly necessary.

With regard to the target audience for ITS education and training, the following categories were among those identified:

- System designers and developers for ITS projects/applications;
- Technicians responsible for system implementation, testing and maintenance;
- Those undertaking innovative R&D in the ITS field;
- Transport specialists acting as advisers to policy makers;
- Transport executives responsible for policy decisions involving ITS;
- Administrative staff involved in assigning and specifying ITS contracts;
- Staff dealing with legal questions associated with the implementation of ITS schemes.

Most courses currently offered at European level are externally funded, which means that participants are not paying market prices. Assuming that financial support in the future is likely to be available only in special cases, it will be necessary to assess the willingness to pay. The development of a realistic business model will be fundamental to the success of a European training and education initiative.

Examples, such as the telematics initiative in Austria, have shown that it is acceptable to charge full course fees when participants and employers are convinced of their value. The web-based courses offered by CITE in the United States also indicate that students are willing to pay for ITS courses which offer credits. One of the ways of reducing the expense of producing new material (and hence the cost to participants) is to create a shared 'pool' of teaching resources. Such a strategy could be adopted in Europe if the major educational institutions involved in ITS teaching were prepared to co-operate within a European network, and if an acceptable 'exchange' mechanism could be established.

7.3.3.3 ITS awareness

An issue frequently raised by representatives of both industry and academia is the general lack of awareness of ITS. It is claimed that the take-up of courses in universities is limited by the low profile of ITS and inadequate information about courses and career prospects. Even within the world of transport, and on the part of transport authorities themselves, inadequate knowledge of the scope and implications of ITS is frequently encountered.

This suggests that pro-active promotion of ITS could help to stimulate greater interest and, by increasing the demand for training, help to raise the level of competence in the profession. A range of dissemination and outreach activities would need to be designed for a wide range of targets, including students (from engineering, and also other faculties) as well as industry, and national, regional and city authorities involved in transport.

7.3.3.4 Distance learning

Despite recognition of the need for raising professional standards in ITS, course participation can be difficult for working people.

One way of obviating the requirement of personal attendance, and hence absence from the workplace, is by adopting one of the many forms of distance learning which have the added advantage of eliminating the travel cost and time, and offering flexibility in the study schedule.

Experience indicates that to be successful, such courses need to be very well designed and organised, and that blended (mixed media) solutions are in general the most effective. This involves a combination of face-to-face teaching and sessions via the Internet, television networks, video, CD ROMs, etc. An approach which has proved effective in the United States for complex subjects, such as ITS Architecture, is for introductory sessions to be taken online and then followed up with traditional classroom teaching.

7.3.3.5 *Language question*

A practical difficulty raised by a centralised approach to ITS education in Europe regards language. While the *de facto* international language is English, not all potential participants (especially those above a certain age!) have sufficient familiarity to easily follow lectures or read documentation. Courses, workshops, and written material produced (and especially if funded) at European level will inevitably be in English with translations arranged locally. Support can be given, for example, through the preparation of multi-lingual glossaries of technical terms.

7.3.3.6 *ITS qualification and accreditation*

The desirability of establishing an internationally recognised ITS qualification (or set of qualifications) has been strongly endorsed throughout Europe. It is felt that this would add status to ITS as a field of expertise and, by encouraging students to take recognised courses, help to raise the standard of training.

Such a move would require a mechanism for the official recognition of courses and a system of 'quality control'. Since existing credit systems vary enormously from one country to another, it is important to support the efforts already being made to establish Europe-wide accreditation procedures. The possibility of being able to 'mix and match' courses offered in different institutions, in the same country or abroad, would not only facilitate the acquisition of a comprehensive grounding in ITS, but also promote an international perspective.

7.3.4 *Future opportunities*

Scope clearly exists for initiatives aimed at improving the quality and availability of ITS training and education. A co-operative approach at the European level would have the advantage of offering scope for optimising the expertise and resources. An arrangement for sharing material would mean that not only the total time and effort but also the overall 'production' costs would be reduced. By making courses, training facilities and general information widely available, students and professionals in all regions of Europe would be able to make better choices regarding the training for their ITS career and have access to high quality instruction.

There are several important long-term strategic benefits which further justify such an initiative:

- Raising the level of professional ITS skills would lead to better designed ITS applications and improve the overall efficiency of transport services, bringing social and environmental benefits;
- Encouraging a trans-national perspective would help to promote harmonisation and compliance with international standards, and support the marketing of European ITS products and services;
- Achieving the rapid and efficient dissemination of research results and best practice would raise the technological excellence and quality of ITS applications in Europe.

To make these objectives a reality requires the creation of a *permanent and financially self-supporting organisation* which would promote a co-operative approach to ITS training and education in Europe. The remit of such an organisation should be:

(a) To improve the quality and availability of ITS training and education for students and researchers;

(b) To provide refresher courses for those already working in the ITS field;

(c) To undertake outreach and promotional activities to raise general awareness of ITS.

7.3.5 Recommendations

The organisational structure which it is felt would best meet the above objectives is a dual structure consisting of a European Network for ITS Training and Education backed up by an Operational Centre. The latter would have the role of providing the Network with practical and administrative support.

The long-term aim should be to extend such initiatives to as many countries as possible, and for the Network itself to become a permanent and financially self-supporting institution. Since some form of funding is nevertheless likely to be required in an interim period. It is therefore recommended that a two-stage strategy should be initiated:

1. The drawing up of a voluntary agreement between interested institutions (to include the major European universities involved in the teaching of ITS-related subjects), e.g. through a Memorandum of Understanding. This group could initiate a series of co-operative activities such as exchanging teaching material, proposing curricula for ITS, and organising open workshops. It should also seek ways of obtaining some financial support (from the European Commission and other sources) for undertaking more ambitious activities and defining a Business Model for the Network.

2. Setting up of the Network and Operational Centre on a permanent legal basis. These would be jointly responsible for the creation of teaching material and organisation of courses, workshops and related activities, which would permit them to be self-sustaining.

It should be noted that a European initiative, the ETNITE project funded by the Leonardo programme, is already working in this direction and should be taken as a valuable starting point.

The characteristics foreseen for the proposed Network and Operational Centre are described below.

European Network for training, education and outreach

Such a body would be university-led, but associate membership would be foreseen for a wide range of other ITS-related organisations, including public authorities, research institutes, ITS organisations, national ministries, private industry, transport operators, etc.

The core (academic) members would be responsible for jointly creating a pool of teaching resources which would be characterised by scientific rigour and neutrality. The involvement of other bodies, such as transport authorities and operators, would have the benefit of facilitating inclusion of field visits to real-life examples of best ITS practice, and also making known their own specific training needs.

Fundamental to the success of such an initiative will of course be the ability to identify a valid Business Model. One possible approach would be for members who contribute to the resource pool to gain an agreed amount of 'credits' giving them the right to have access to other material. Contributions could be in various forms: by providing course material, undertaking scientific and technical reviews, editing texts, etc. for which others would pay a fee. This approach would depend, among other things, on copyright and intellectual ownership issues being resolved.

In parallel to this kind of internal exchange of resources, the network could earn revenue and extend the scope of its activities by producing courses for external participants. Among the key targets would be public authorities.

It should be stressed that the basic purpose of the network would be the creation and sharing of resources, not to undertake regular teaching. The exact nature and scope of the activities would be decided by the members, but would be likely to include:

- Producing guidelines on the definition of ITS curricula;

– Definition of mechanism for making available research results
– Providing a structure for e-learning;
– Organisation of short (1-2 days) seminars on key ITS topics

Operational Centre

The role of the Operational Centre would be to serve as a 'Back Office' to the Network. It would carry out support activities, including secretarial duties and administrative and legal functions (e.g. dealing with membership issues, course registration and payment, legal constitution of the network, establishment of ownership of intellectual rights).

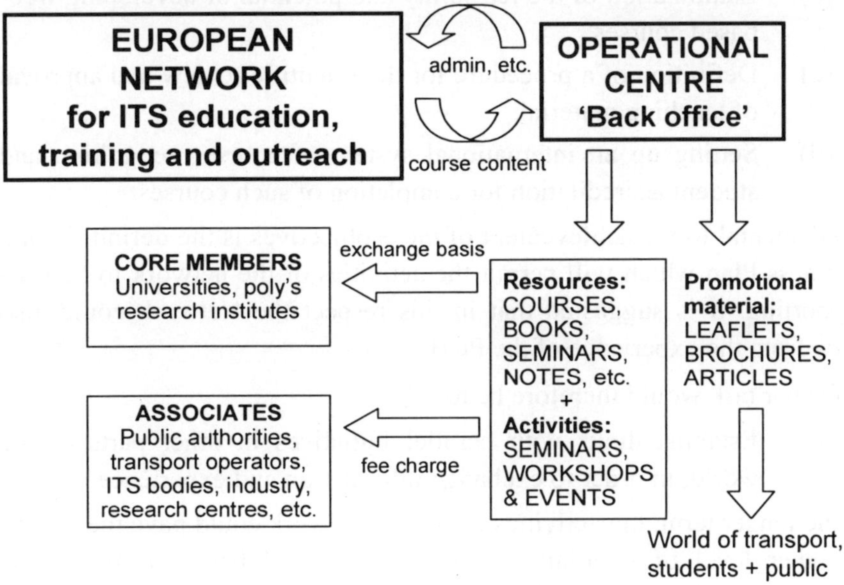

Figure 17: Proposed organisational structure

An important task of the Centre would be an editorial function involving preparation of the teaching material (modules, textbooks, documentation, etc.) in a suitable form for publication, adopting an accepted common format. The content itself would be provided principally by network core academic members (i.e. universities). The Centre would also oversee the production of outreach and dissemination material (brochures, leaflets, etc.) on the basis of proposals received from the network. It is possible that a further responsibility would be the maintenance of a website.

It is conceivable that these functions could be undertaken by a single organisation, or divided between several different ones.

Among the tasks which will need to be tackled jointly by the Network and Operational Centre are:

(a) The creation of a centralised source of information covering educational curricula and training courses available in Europe, seminars, events, etc. (An appropriate model could be the website run by the PCB in the United States);

(b) Examination of the feasibility and potential of developing web-based courses;

(c) Definition of a procedure for the scientific review and approval of teaching material;

(d) Setting up an international system of course recognition and student accreditation for completion of such courses.

Fundamental to the achievement of these objectives is the definition of a Business Plan which will permit the activities of the network to be self-supporting. It is suggested that in this respect the Network could also draw from the experience of the PCB.

A further task would therefore be to:

(e) Establish links with parallel initiatives in other parts of the world, in order to exchange information and experience.

In the longer term, the activities of such a Network could pave the way to a higher level of international co-operation and harmonisation. It is conceivable that the initiative could move towards a recognised European programme for ITS education and training, with recommended curricula for the various areas of study. This would operate in conjunction with an international system of credits awarded for the completion of the courses. (It is important to note that respect for the principle of subsidiarity means that this development would have to be based on voluntary compliance.)

Chapter 8

Conclusions and Recommendations

Intelligent Transport Systems and services are now an integral part of our personal travel and the way in which goods are moved. Whether or not we are aware of it, some or all of the location, identification, charging, communication, management, control and information functions which constitute ITS are used to our advantage when we travel. When we use public transport, ITS applications such as smart cards, internet ticketing or online information screens are often supported by sophisticated management systems for online decisions and priority allocation. When we drive, systems such as ABS, cruise control and navigation make our driving environment safer and more comfortable, and safe behaviour is reinforced with on-road systems such as red-light cameras. Goods are moved with increasing levels of tracking and tracing to improve security and delivery.

ITS systems and services are introduced by either policy or market drivers, which in some situations may coincide. Policy drivers relate to safety, efficiency, the environment, social goals and the need to support a healthy economy. The main approach to the latter is by trying to provide transport infrastructure and facilities which support a competitive European market. Whilst ITS is an essential part of this process, governmental support for ITS research, development and deployment more generally helps to provide a strong home base from which European industry can market worldwide. Market drivers are often more straightforward. A product must be sufficiently attractive for it to be purchased at a price and in sufficient numbers for a profit to be generated. It may or may not contribute to any of the governmental policy targets, but governments/EC can regulate to control standards and use. This is not always effective because the decision processes of industry and those of government can be very different in character and timing. An example of this problem has been the widespread use of mobile telephones whilst driving.

Thus, whilst much ITS development has taken place, the rate of deployment has been slower than had been expected and many of the potential policy and market gains have not been achieved. This book has addressed this issue in a series of key ITS areas and detailed discussions and recommendations have been developed in the preceding chapters, which are self-standing. At the heart of the findings and recommendations is the understanding that stakeholders, particularly public and private bodies, have not come together in a sufficiently meaningful way to enable integrated ITS developments to take place. Developments have been fragmented and whilst each has been successful in its own right, the major benefits of coherent integrated systems and services have not occurred. Key findings and recommendations which cut across application areas are given below:

(a) Generally, the development of vehicle-based Intelligent Transport Systems (ITS) and services is market led and end users will determine which systems and services will succeed and which will fail. However, in some circumstances the role of governmental bodies is crucial in acting as a catalyst for complex initiatives, providing supporting evidence to potential commercial/industrial leaders, and raising awareness of the need for and benefits to be gained from Intelligent Transport Systems. Examples of successful governmental intervention include developments in system architecture which have enabled clearer product specification, and the user acceptance of seat-belt legislation through government promotion of the safety implications. It is strongly recommended that dialogue between governmental and industrial bodies is focused on identifying and pursuing those actions where government intervention can act as a catalyst to market development of ITS products and services which meet their policy objectives. Too often, government responses are neither timely nor appropriate.

(b) The introduction of new technology will usually require substantial financial commitments. Determining clear roles, responsibilities and commitments of the various stakeholders can be particularly difficult and has led to a delay in deployment of many services, particularly those which cannot be introduced in an incremental or evolutionary way. It is recommended that clear future visions are developed and agreed by all stakeholders with common 'ownership" and commitment. These will include deployment paths to revolutionary applications where this is appropriate. (The setting up of the European Road Transport Research Advisory Council is a step in the right direction, if meaningful cross stakeholder dialogue is achieved.)

(c) The national characteristics, physical environment, levels of development and economic activity vary substantially between countries and regions of the EC. These differences have increased particularly since the recent expansion of membership from 15 to 25. There is no single infrastructure-based transport solution across Europe, ITS or otherwise, and subsidiarity remains an important aspect for ITS/transport research, as for other issues. It is recommended that further research is undertaken to better understand which key features of national and regional infrastructure-based transport applications should be common and the extent of such commonalities.

(d) The introduction of many intelligent transport systems and services involves the identification and location of individuals or vehicles, and may also involve the transfer of money. This has implications for privacy and security. For many systems, individuals may have a choice as to whether or not to use them, and it is likely that considerations of any loss in privacy will be more than offset by the benefits to be gained. However, for some services, such as road user charging, there may be no practical way of avoiding some loss in privacy. Also, system security cannot be guaranteed but will continue to develop to meet evolving threats. Probably, the key to user confidence is to generate an atmosphere of trust. This will require integrated research, policies, and awareness actions. In particular, many technologies may be applied in such a way as to best protect the privacy and security of system users. It is recommended that research be undertaken to understand how an acceptable level of privacy can be achieved with ITS systems, whilst retaining the individual benefits. The balance between privacy and security will be a growing issue.

(e) Whilst technologies continue to mature, their deployment, and hence the benefits of deployment, has often been extremely slow. Indeed, only technologies which have been evolutionary in nature have been taken up, whilst those which have been more revolutionary have not, even those for which the potential benefits are much greater. This may be the result of several factors, such as a lack of convincing research evidence, the lack of conviction by stakeholders, or conflicts of those with vested interests. However, at the core, the lack of a clear vision of the future role of technology in transport to which everyone can ascribe continues to be a substantial restraint on development and deployment. It is recommended that such a vision should be developed with 'ownership' by technical, administrative and political bodies in a way which transcends the more normal timescales of decision processes and commitments. This will require a process to research a policy framework for ITS applications.

(f) Intelligent transport systems and services can enable travellers to make better journey choices, increase capacity for both public and private networks, improve reliability and safety, reduce delay and increase accessibility. They can also be used for control, enforcement and charging. They may be applied in ways which enable existing policy objectives to be met or new ones to be set. However, they will generally not lead to substantial increases in capacity, and a sustainable future will either require a reduction in demand at certain times and in certain places or new infrastructure to be built. It is recommended that research is undertaken to better understand the impact and potential of ITS in the broader context of land use and transport infrastructure decisions.

(g) The training of practitioners and researchers is necessary to support the development and deployment of ITS. Many of the ITS applications require specific subject area skills, such as electronic engineering, but there is an increasing demand for the enabling skills for deployment related to areas such as architecture or policy, which interface the traditional disciplines. It is recommended that new training schemes are put in place to develop the necessary European-wide skills base in ITS.

(h) Delay in the development and deployment of ITS will not only result in a lack of benefits in the short term, but will also enable product development outside the EC to enter the market first. In such circumstances the benefits to the EC will be substantially reduced. The EC still leads in many ITS areas, but major initiatives in the USA and Japan and growing activities in areas such as China and India mean that more research and development is needed to maintain the EC world position. It is recommended that substantial research and development funding is allocated to meet these overall recommendations and those throughout this book. For example, the European lead in integrated traffic management and control needs support to incorporate the developing new ITS opportunities and the related novel and radical policy approaches to the use of road space.

(i) The acceptance of ITS will be greatly enhanced by the development of interfaces with users which match their individual needs more exactly. Users must be able to access services in an easy and secure way. It is recommended that research and development should be undertaken on the ways in which ITS systems and services interface with people. This relates not just to HMI issues, but also to the underlying structures of the systems and services themselves.

(j) New opportunities will develop which may radically change the potential of new services. For example, nanotechnologies may be used to gather information. It is recommended that such new developments must be monitored and adapted where there is benefit, and system architectures must allow for flexibility.

Recommendations have been made throughout this book to take forward a wide range of ITS technologies to meet the needs and desires of society issues and of the market. Success will only be fully achieved with a co-ordinated approach which will enable the inevitably limited resources to be focused on research, demonstrations and delivery in a coherent manner by all stakeholders. This is the task which faces the EC in the coming years.

Appendix A

Research Projects related to ITS

This section gives an overview of projects related to the topics discussed in this document. The projects have been grouped according to their main focus, although frequently several issues were dealt with in the same project.

3. TRAVELLER SERVICES

3.1 Transport Services		
Project	Programme	Contents
Intermodality – integration of modes		
CIVITAS I & CIVITAS II	5th & 6th FP - TREN	The CIVITAS initiative aims to achieve a significant change in the modal split towards sustainable transport modes. This objective is carried out through the combination of technology and policy based strategies. Eight measure areas have been identified as the basic building blocks.
METEOR	5th FP -TREN	Independent parallel project to compare and assess the results of the CIVITAS I projects in a harmonised way across all sites (CIVITAS I project).
MIRACLES	5th FP -TREN	Combination of innovation, technology and policies with the support of communication media so that with the active participation of citizens, traffic, energy consumption, noise and air pollution can be reduced (CIVITAS I project).

MUSIC	4th FP – TREN	Development, demonstration and assessment of novel methods of traffic control and management as a cost-effective means to reduce congestion, improve the efficiency/cleanliness of urban travel and influence modal choice.
TAPESTRY	5th FP – GROWTH	Assessment of the effectiveness of travel awareness, publicity, information and education initiatives in encouraging travellers and goods operators to adopt a more sustainable, intermodal travel behaviour. This is accompanied by recommendations, guidance and practical advice on the potential of multimodal travel awareness campaigns and their cost effectiveness.
TELLUS	5th FP –TREN	Increasing the modal share in favour of public transport. This is carried out through improvement of intra-organisational co-operation at city level and improvement of public private co-operation (CIVITAS I project).
TRENDSETTER	5th FP –TREN	Amelioration of urban air quality, noise levels and congestion while supporting exceptional mobility and urban quality of life i.e. through advanced mobility management schemes or improved goods logistics and increased use of low emission vehicles (CIVITAS I project).
VIVALDI	5th FP –TREN	Demonstrating an integrated package of innovative transport strategies and measures and to assess their contribution to improve urban vitality and economic success, social inclusion, health and well-being of the citizens and sustainability (CIVITAS I project).

Passenger transport systems and operation		
AUSIAS	4th FP – TAP	Demonstration of advanced transport telematics in urban sites with integration and standardisation, e.g. dynamic bus scheduling and remote maintenance monitoring systems.
BERTA	National (Germany)	Basic tool for maintenance of all ITS data for bus, tram and metro, i.e. drawing of timetables, monitoring of actual operations, supervising safety and security.

EUROMAIN	5th FP – IST	European railway open maintenance system which will allow remote monitoring and diagnosis of complex systems.
EUROPE-TRIS	5th FP - IST	Developing tools to address the planning process in railway companies. The activities that are supported include infrastructure management and transport operations.
ISCOM	5th FP - IST	Information systems for combined mobility management in urban and regional areas – development and demonstration of multimodal transport information and services.
PRISCILLA	5th FP - IST	Bus priority strategies and impact scenarios developed on a large urban area – taking up of existing bus priority practices and improving them for wide networks. Evaluation of main impacts, demonstration of benefits and dissemination of best practice for adoption.
QUARTET PLUS	4th FP - TAP	Validation of European urban and regional integrated road transport environments based on open system architecture.
TABASCO	4th FP - TAP	Validation of prototypes of enhanced traffic control and traveller information systems, i.e. traffic-adaptive signal control system, hazard warning system, expert system for traffic incident management.
TRASCOM	5th FP - IST	Aims at developing multi-media applications using mobile internet telephony to improve user mobility, combining all modes of transport.
TRIDENT	5th FP - IST	Aims to establish mechanisms for the sharing and exchange of common and reusable data to enable and support multimodal services.

Payment Systems		
ADEPT/ ADEPT II	4th FP - TAP	Using an intelligent transponder and smart card for a multitude of automatic debiting, electronic payment and other complementary road/rail transport informatics applications.
CALYPSO	4th FP - TAP	Contact and contactless environments yielding a citizen pass integrating urban services and financial operations.

DISTINCT	4th FP – TAP	Deployment and integration of smart card technology and information networks for cross-sector telematics.
OMNIPURSE	5th FP – IST	Enable contactless e-purse payments on a smart card by developing the necessary technical infrastructure to allow for secure, high speed financial transactions to ensure payment for transport facilities.
SM-PAYSOC	5th FP – IST	Secure mobile payments and services on chip – development of standards, systems and rules to ensure the widespread interoperability of multi-application smart cards.
TELEPAY	5th FP – IST	Development of an innovative telepayment system for multimodal transport services using portable phones (SMS and WAP).
TR@VELSMART	5th FP – IST	Business concept for an intelligent smart card and Internet-based customer relationship management service for European tourism destinations.
TRIANGLE	5th FP – IST	Proof of concept for a simple, workable and manageable interoperable solution for door-to-door travel based on chip-card.

Demand Responsive Transport		
FAMS	5th FP – IST	Implementation and trial of a flexible agency for collective, demand-responsive mobility services. Evaluation of the viability and impacts in real business cases and gathering knowledge and best practice.
IGT	National (UK)	Intelligent Grouping Transportation – passengers with compatible itineraries are placed onto the same taxibus. Booking, co-ordination/location of vehicles, etc. is realised through a computerised system.
INVETE	5th FP – IST	Specification development and validation of a modular intelligent in-vehicle terminal (IVT) for multimodal, flexible collective transport services.
SAMPLUS	4th FP – TAP	Extended evaluation of telematics technologies used in Demand Responsive Transport (SAMPO).

SAMPO	4th FP – TAP	Systems for advanced management of public transport operations – evaluating the potential and functionality conferred by the application of data communication technologies to Demand Responsive Transport Services.
SCRIPT	4th FP – TELEMATIC S 2C	Feasibility study on utilising telematics processes to meet the needs of the rural communities by enhancing access to services and reliable information.
VIRGIL	4th FP – TRANSPORT	Verification and improvement of the access to services and transport in rural areas, by identifying and disseminating good practices and experiences on rural access to transport, with emphasis on door-to-door solutions for the transport of people and goods.

3.2 Information Services		
Multimodal information		
EuroSPIN	4th FP – TAP	Acquisition of seamless multimodal public transport information, development of a system capable of producing arrival/departure times, connecting services and journey plans for various transport modes.
EU-SPIRIT	4th FP- TREN	Creation of a system for providing seamless timetable information throughout Europe.
IM@GINE IT	6th FP – IST	(Part of the eSafety initiative) is combining previous results and agent-based technology to create a universal platform covering urban, interurban, and cross-border areas. This platform will serve as a single access point through which the end user can obtain a set of TTI services everywhere in Europe.
INFOTEN	4th FP – TAP	Focus on language-independent systems for traffic information exchange, multimodal traveller information services and advanced driver warning systems for the Alpine area and Central Europe.

ODIN	5th FP – IST	Development of innovative paradigms for the design of open, distributed and networked tools for TTI with the aim of favouring a new class of just-in-time, interactive, value-added, map-based and personalised services for tourists, business people and commuters. The project defined a global system architecture for a mobile delivery platform and services compliant with the ISO/ODP (Open Distributed Processing) and prototyped an integrated toolkit to access information sources using XML/Java and view them on wireless hand-held devices.
TITAN	4th FP – TAP	Validation of a European reference data model for public transport operations (Transmodel).
TRIDENT	5th FP – IST	Aimed to establish common mechanisms for exchanging data between content owners for different modes (bus/tram/metro, rail and road). It also investigated the organisational and strategic issues hampering intermodal travel.

Telematics platforms		
ENTERPRICE	4th FP – TAP	Developed a Traffic Information Centre (TIC) able to generate multimodal traffic information from heterogeneous raw data. This was 'fused' and presented in a form suitable for use by various private and public information services.
ISCOM	4th FP –TAP	Development, demonstration and validation of electronic timetable information and the creation of 'mobility centres' in cross-border regions (Alsace and Baden Württemberg) and two European capitals with mass transport problems (Rome and Vienna). Value-added services on digital communication networks are provided both to transport agencies and road users (information services can be accessed either directly by users or through staff in the mobility centres).
SMITH	4th FP – TAP	Developed the TITOS platform in Turin, Italy, which was funded jointly by the EC, the Turin municipality and city transport authority. It is still in operation and provides information on city traffic, public transport and parking over a large number of different channels.

User requirements		
IADS	4th FP – TAP	Organised large scale demonstrations of information services.
INFOPOLIS 2	4th FP – TAP	Examined travel as a dynamic process in which the user carries out tasks within three contexts: pre-trip planning, on-trip (tracking) and end-trip (assessment). It concluded that the main role of information must be to reduce the uncertainties inherent in travelling.
INFOVILLE	4th FP - TAP	Large-scale real-life demonstration and evaluation of user-driven cross-sector telematics supported by a standardised integration platform. Also examined the issue of privacy raised by 'user profiles'.
TELSCAN	4th FP - TAP	Aimed to ensure that developers of advanced transport telematics take into consideration the specific needs of elderly and disabled travellers and drivers.
TR@VELSMART	5th FP - IST	Developed an innovative loyalty programme for regional tourism based on the use of smart cards and Internet-based services.

Communication Technologies and Protocols A number of research projects in current and past programmes have focused on the exploitation of new technologies for TTI (e.g. GSM, DAB, Internet, microwave links) through emerging and consolidated protocols such as WAP, XML, TCP/IP, TPEG and DSRC.		
ATLANTIC	5th FP – DG INFSO	A thematic long-term approach to networking for the telematics and ITS community. Survey of transport and traveller information services in Europe.
DIAMOND	5th FP - IST	Examined a wide range of DAB based applications - from simple traffic information to dynamic navigation tasks - and the technical and commercial feasibility of integrating ITS services over the DAB channel with technologies such as GSM and GPS.
ENTERPRICE/ EUROSCOPE	4th FP -IST	DAB trials were carried out in ENTERPRICE (in Hessen, Germany) and EUROSCOPE (Cologne).

ITSWAP	5th FP - IST	Evaluated the technical and commercial feasibility of offering services over the Wireless Application Protocol (WAP), the emerging industry standard protocol for GSM (suited to mobile Internet applications using the GSM channel).
PRETIO	EC e-TEN	Market validation of information services over hybrid communications systems. DAB and mobile telephony are being used to deliver multimedia services in three test areas, offering different levels of services, from a basic version to a customised 'pay per use' service.
PROMISE	5th FP - IST	Development of GSM-based information services.
TPEG	5th FP - IST	Investigated exploitation of the TPEG independent protocol (for DAB and Internet) for the transmission of traffic and travel information within digital broadcast systems, with a view to standardisation.

Human-Machine Interface (HMI)		
CATCH-2004	5th FP - IST	Developed a multilingual conversational system with a unifying architecture across devices and services. The aim was to provide multilingual vocal access to applications and information sources on the Internet (public and private service providers) by mean of devices such as kiosks, standard telephones and smart wireless devices.
COMUNICAR	5th FP - IST	Developed an on-board multimedia interface designed to harmonise the message input from the driving support systems, telematics services, and functions regarding driver comfort and convenience.
CORETEX	5th FP - IST	Worked on speech recognition technology to make it less sensitive to linguistic factors, and more suitable for European languages.
DIAMOND	5th FP - IST	Considered user requirements and the HMI implications of WAP and DAB based services.
INFOPOLIS	4th FP- TAP	Developed guidelines for the design of TTI systems, including indications for content, layout, icons, presentation, ergonomics, sequencing of information, etc.
ITSWAP	5th FP - IST	Considered user requirements and the HMI implications of WAP and DAB based services.

PEPTRAN	5th FP - IST	Carried out experiments with hand-held devices and a car navigation system to guide a user from point to point within a city.
TRAVEL-GUIDE	5th FP - TREN	Developed guidelines for TTI integrated with traffic management systems (using in-vehicle information devices and road side-systems). It assessed the needs of drivers in terms of content, presentation, availability, reliability, and the timing and priority of the information. Tests were conducted and new methods suggested to meet requirements raised by the development of the Trans-European Networks.

4. VEHICLES AND INFRASTRUCTURE

4.1 Advanced Driver Assistance Systems		
ADASE 2	5th FP – IST	About 30 projects organised in the ADASE cluster, doing research for driver assistance systems and related activities to improve road transport safety, efficiency and comfort.
ADVISORS	5th FP – IST	Analysis and assessment of appropriate ADAS in a multidisciplinary approach, in order to gain new policy insights. The approach led to an integrated framework suited for designing road safety policies to help realise suitable ADAS.
EUCLIDE	5th FP – IST	Development of a reliable integrated driver assistance support system based on radar and far infrared sensors, to monitor the area ahead of the driver to provide an effective support, especially in cases of night and adverse weather conditions.
PROMETHEUS	DG XII	Programme for European Traffic with Highest Efficiency and Unprecedented Safety, encompassing the areas of improved driver information, active driver support, co-operative driving, and traffic and fleet management.
STARDUST	5th FP – IST	Investigations into the various aspects and impacts of the deployment of ADAS/AVG technologies that could be envisaged to 2010 for 3 European cities, including state-of-the-art review, elaboration of deployment scenarios, market assessment, human factors studies, impacts simulation and evaluation.

4.2 Co-operative Vehicle Highway Systems		
CHVS	National (UK)	Development of outline business cases for Co-operative Vehicle Highway Systems.
IVI	National (USA)	Aims to advance the safety, efficiency and security of surface transportation.
PROMETHEUS	DG XII	Programme for European Traffic with Highest Efficiency and Unprecedented Safety, encompassing the areas of improved driver information, active driver support, co-operative driving, and traffic and fleet management.
VII	National (USA)	Aims to create an enabling communications infrastructure to support a spectrum of public authority and vehicle manufacturer interests.

4.3 Human Machine Interaction		
ADASE 2	5th FP – IST	About 30 projects organised in the ADASE cluster, doing research for driver assistance systems and related activities to improve road transport safety, efficiency and comfort.
ADVISORS	5th FP – IST	Analysis and assessment of appropriate ADAS in a multidisciplinary approach, in order to gain new policy insights. The approach led to an integrated framework suited for designing road safety policies to help realise suitable ADAS.
COMUNICAR	5th FP – IST	Development of two vehicle demonstrators (a city car and an upper class car) integrating a new concept of multimedia HMI able to harmonise the messages coming both from the traditional on-vehicle information (e.g. tachometer) and the new functions for driving, like telematics services or other information concerning comfort and entertainment.
EDEL	5th FP – IST	Development of an advanced vision enhancement system for night vision applications based on near infrared sensor, a novel illumination system and an adaptive human machine interface to reduce the number of road accidents in Europe.
EUCLIDE	5th FP – IST	Development of a reliable integrated driver assistance support system based on radar and far infrared sensors, to monitor the area ahead of the driver to provide an effective support, especially in cases of night and adverse weather conditions.

E-MERGE	5th FP – IST	Development of an in-vehicle emergency call solution to ensure that a manual or automatic call for assistance arrives at the PSAP (Public Safety Answering Point), and that proper actions can consequently be taken to dispatch assistance to the vehicle.
HASTE	5th FP – IST	Exploration of the relationships between task load and risk in the context of safety-critical driving scenarios. This includes consideration of issues of workload and situation awareness, and the identification of the best indicators of risk.
ITSWAP	5th FP – IST	Development of innovative services that deliver Internet-like information to mobile users. To achieve this, the project will establish the technical and commercial feasibility of Intelligent Transport Systems (ITS) services provided over Wireless Application Protocol (WAP).
ROADSENSE	5th FP – IST	Support of the European motor vehicle manufacturers in developing industry codes of practice in the field of vehicle safety and especially in Human Vehicle Interactions (HVI), by delivering guidelines for the methods of HVI tests and the development a driver behavioural simulation environment to evaluate new technology proposals.
STARDUST	5th FP – IST	Investigations into the various aspects and impacts of the deployment of ADAS/AVG technologies that could be envisaged to 2010 for 3 European cities, including state-of-the-art review, elaboration of deployment scenarios, market assessment, human factors studies, impacts simulation and evaluation.

4.4 Emergency Response		
AIDER	5th FP – IST	Development of on-board equipment providing automatic incident detection, and communication with emergency centres.
CGALIES	5th FP –IST	Definition of the requirements for a Pan European common location provisioning mechanism, that can be accessed and used by the European 112 community and emergency service operators .

E-MERGE	5th FP – IST	Development of an in-vehicle emergency call solution to ensure that a manual or automatic call for assistance arrives at the PSAP (Public Safety Answering Point), and that proper actions can consequently be taken to dispatch assistance to the vehicle.
LOCUS	5th FP – IST	Support and expertise regarding the definition of location based emergency services.
RESCUE	6th FP – IST	Focus on the automatic assessment of the type of emergency and forwarding information to appropriate centres.

4.5 Enforcement in ITS		
VERA	4th FP – TAP	VERA explored issues and opportunities arising through the use of digital systems for enforcing road traffic laws and regulations. The project made recommendations on the use of digital systems for automating the enforcement process and for enforcing violators across national borders. Enforcement systems' trials and implementations.
VERA2	5th FP – TREN	Project to support pan-European cross-border automatic enforcement by video and define relationships between enforcement agencies to govern the use of these tools.

5. NETWORK MANAGEMENT

5.1 Traffic Management and Control		
Traffic Management and Control		
CIVITAS I & CIVITAS II	5th & 6th FP – TREN	The CIVITAS initiative aims to achieve a significant change in the modal split towards sustainable transport modes. This objective is carried out through the combination of technology and policy based strategies. Eight measure areas have been identified as the basic building blocks.
DRIVE 2	3rd FP – FRAMEWORK 3C	Community Programme. Includes seven areas of major operational interest: demand management, travel and traffic information systems, integrated urban traffic management systems, integrated inter-urban traffic management systems, driver assistance and co-operative driving, freight and fleet management and public transport management.
EURAMP	6th FP – IST	Collective European action focused on ramp metering control measures on European motorways with the aim of improving safety and increasing efficiency of traffic flow.
LLAMD	3rd FP – DRIVE 2	Development and demonstration of aspects of the integration of Advanced Transport Telematics (ATT) systems within an integrated road transport environment. LLAMD includes different sub-projects dealing with urban traffic control/ management, park-and-ride/parking/public transport information, dynamic route guidance systems, fleet management and safety aspects.
PRIME	5th FP – IST	The project aims to increase the effectiveness of incident detection and incident management on motorways and adjacent urban networks and to increase road safety, through the development of innovative methods, building on recent achievements in related EU projects.

QUARTET	3rd FP – DRIVE 2	Development of a fully Integrated Road Transport Environment (IRTE) based on five integrated Advanced Transport Telematics (ATT) technologies – target IRTE architectures, Environment Control, Dual Mode Route Guidance, Public Transport Management and Information Systems and Emergency Call.
QUARTET PLUS	4th FP – TELEMATICS 2C	Validation of European urban and regional integrated road transport environments (see also QUARTET) based on open system architecture.
TAPESTRY	5th FP – GROWTH	Assessment of the effectiveness of travel awareness, publicity, information and education initiatives in encouraging travellers and goods operators to adopt a more sustainable, intermodal travel behaviour. This is accompanied by recommendations, guidance and practical advice on the potential of multimodal travel awareness campaigns and their cost effectiveness.
UTOPIA	5th FP – IST	To provide decision-makers with the necessary information base, tools, methods and guidelines for hastening the market introduction of the most appropriate transport solutions based on new propulsion technologies and systems (NPS).

Urban Traffic Management and Control		
EUROCOR	3rd FP – DRIVE 2	Development and implementation of models and online control strategies for the effective management of traffic in urban corridors (integrated urban networks and motorways), including traffic signals, ramp metering and Variable Message Signs.
HIPERTRANS	4th FP – TRANSPORT	Development of an accurate, high-speed, and visually interactive simulator for a road transportation network within a high-performance computing (HPC) environment that is capable of interfacing with urban traffic control (UTC) systems in real time.
ICAROS NET	5th FP – IST	Development and implementation of a networked interactive computational environment that allows the minimisation of uncertainty in decision making regarding operational air pollution control and abatement in the urban environment and enhances the coherence in trans-boundary environmental monitoring.

INCOME	4th FP – TRANSPORT	Development, integration and evaluation of strategies/software for optimisation of Urban Traffic Control (UTC), Driver Information Systems (DIS) and Public Transport Systems (PTS) within Urban Traffic Management Systems (UTMS).
INTRAMUROS	4th FP – TRANSPORT	Provision of new knowledge on the integration of the different actors involved in urban transport management systems, including market oriented urban transport systems, and to develop a conceptual methodology for assessing the integration of these actors.
MUSIC	4th FP – TRANSPORT	Development, demonstration and assessment of novel methods of traffic control and management as a cost-effective manner to reduce congestion, improve the efficiency/cleanliness of urban travel and influence modal choice.
PRISCILLA	5th FP – IST	Bus priority strategies and impact scenarios developed on a large urban area – taking existing bus priority practices and improving them for wide networks. Evaluation of main impacts, demonstration of benefits, and dissemination of best practice for adoption.
SMART NETS	5th FP – IST	Aims at enabling a significant improvement of the international state-of-the-art in real-time network-wide urban traffic control via application, demonstration, and comparative evaluation of the new-generation control strategy TUC (Traffic-responsive Urban Control).
TABASCO	4th FP – TELEMATICS 2C	Demonstration project implementing multimodal information and control systems to solve transport problems of cities and regions. User-oriented validation of transport telematics systems and their integration in cities and in their surrounding regions to produce a more efficient transport system as a whole.

Traffic Management and Control – Monitoring		
EFFECT	4th FP – TELEMATICS 2C	To predict poor local air quality in real time and then to initiate effective traffic demand management strategies/measures to reduce pollution levels in particular problem areas.

HEAVEN	5th FP – RTD	To develop and demonstrate a fully integrated decision support system for a healthier environment through real-time monitoring and modelling of key pollution sources. Establishment of a data platform for the assessment of emissions and health effects of air pollutants and noise caused by traffic.
ICAROS NET	5th FP – RTD	Development and demonstration of an integrated computational assessment of urban air quality via remote observation systems network.

Traffic Management and Control - Freight		
EDRUL	5th FP – IST	Investigation, development and validation of an innovative IST platform and the supported service/business models, for improved management of freight distribution and logistics processes in urban areas. The project is based on eCommerce/eBusiness architectures and concepts and supports demand-responsive logistics schemes.
LEAN	4th FP – TRANSPORT	Development and demonstration of new concepts, telematics applications and administrational tools to distribute and collect goods in urban areas. The project included alternative transport mode recommendations to support significant modal shift to rail.
MOSCA	5th FP – IST	Key project objective is to provide a set of tools for improving efficiency of door-to-door transport of goods in urban areas, by collaboratively providing demand-and-supply side information in one single environment/system.

Interurban Traffic Management and Control		
DACCORD	4th FP - TELEMATICS 2C	Development and validation of a telematics-based system for co-ordinated traffic control on motorways linking urban areas, based on an open system architecture for similar purposes.

5.2 Road User Charging		
CUPID	5th FP - TREN	A thematic network for disseminating state of the art information about urban transport pricing and best practice based on the PRoGREss demonstrations.
DESIRE	5th FP- TREN	A research project designed to assess, through the development of realistic case studies, the prospects for inter-urban road pricing in Europe.
EUROPRICE	5th FP - TREN	A project which fosters debate and support at the political level and provides a focus for city/regional issues in the debate.
IMPRINT	5th FP -TREN	A thematic network which sets out to promote the implementation of fair and efficient transport pricing.
PROGRESS	5th FP - TREN	A demonstration and research project into road pricing in eight cities across Europe.
RCI	6th FP- TREN	Development of an open framework enabling road charging interoperability at a technical level.

5.3 Road and Traffic Monitoring		
APOLLO	5th FP – IST	Creation of an intelligent tyre for improving road traffic safety – integration of innovative sensors into tyres for monitoring tyre condition, road condition and tyre-road interaction.
CARSENSE	5th FP - IST	Sensing of car environment at low speed driving (radar, laser, stereo video camera).
EFFECT	4th FP - TELEMATICS 2C	To predict poor local air quality in real time and then to initiate effective traffic demand management strategies/measures to reduce pollution levels in particular problem areas.
GALILEO	EC	European Satellite Navigation System. The system will provide quality services in all environments: cities, indoors and outdoors, land, sea, air, etc. in synergy with other service provisions, such as telecommunication, surveillance or observation.
HEAVEN	5th FP – RTD	To develop and demonstrate a fully integrated decision support system for a healthier environment through real-time monitoring and modelling of key pollution sources. Establishment of a data platform for the assessment of emissions and health effects of air pollutants and noise caused by traffic.

ICAROS NET	5th FP – IST	Development and implementation of a networked interactive computational environment that allows the minimisation of uncertainty in decision-making regarding operational air pollution control and abatement in the urban environment and enhances the coherence in trans-boundary environmental monitoring.
SMART DUST	National (USA)	Autonomous sensing and communication in a cubic millimetre – self-contained, millimetre-scale sensing and communication platform for a massively distributed sensor network.
TEMPO	5th FP - TREN	Programme to stimulate a harmonised and synchronised deployment of intelligent transport systems and services on the Trans-European Road Network – realised by 6 Euro-regional projects ARTS, CENTRICO, CORVETTE, SERTI, STREETWISE, VIKING.
VIKING	5th FP - TREN	Co-ordination of traffic management schemes and implementation of Intelligent Transport Systems in Scandinavia and five regions in northern Germany (TEMPO project).

6. GOODS TRANSPORT

6.1 Long Distance Freight		
BRAVO	6th FP - TREN	Aims to increase volume of rail freight.
CREATING	6th FP	Concepts to Reduce Environmental impact and Attain optimal Transport performance by Inland Navigation. ICT for transhipments to improve efficiency.
INTERMOD-TRANS	6th FP	Compatibility of transhipment equipment and technologies to increase terminal capacity.
RE-ORIENT	6th FP -TREN	Best practice pilot for road/rail transhipments.
VISIONS	6th FP	Interfaces for information systems for trucks to permit communication, via onboard devices, with external systems. Setting up of pilot system.

Policy and Research Strategies		
ERTRAC	6th FP - TREN	Advisory Council for future research with a focus on breakthrough technologies.
FILIER	6th FP -TREN	To provide the transport and logistic companies with information about organisational and technological innovations, to verify the impact of the telematics and the new organisational modalities in the freight transport and logistic sectors.
FUNDING	6th FP - TREN	Guidelines for the rational funding of transport infrastructure, including charging schemes.
GRACE	6th FP - TREN	Research on accounts and cost estimation.
IMPRINT-NET	6th FP - TREN	Discussion platform for analysis of the costs of infrastructure use.
LOG-BASED	6th FP - RESEARCH	Development of logistics based design process.
NEW OPERA	6th FP - TREN	Definition of strategy to increase rail freight.
RCIPP	6th FP - TREN	European interoperability of tolling technologies and procedures.

Intermodality		
EUTP	5th FP - GROWTH	To create a permanent and dynamic network: to enhance exchange of data and information; and to create synergy in the European research effort related to intermodal freight transfer points.
FV-2000	4th FP	Development of user-oriented guidelines and simulation tools for the evaluation of the Freight Village structure and organisation in order to increase the attractiveness of intermodal transport for industrial and transport operators.
IRIS	4th FP	To demonstrate the feasibility of intermodal transport over short and medium distances and to establish the factors which make this kind of transport a success.
SPIN	5th FP - GROWTH	To develop a tool which can identify opportunities and barriers in achieving a modal shift in freight transport.
THINK-UP	5th FP – GROWTH	To define a common understanding of the core and competitive market segments for the different transport modes in order to appraise their core and competitive position.

6.2 Urban Delivery		
BESTUFS, BESTUFS II	5th, 6th FP	Forum to gather user needs from all stakeholders and to promote dissemination of best practice for urban freight delivery.
CIVITAS II	6th FP -TREN	Group of projects promoting sustainable urban transport. Some include goods transport aspects (CARAVEL, MOBILIS, SMILE, SUCCESS).
E-THEMATIC	5th FP	Examines implications of e-fulfilment for online ordering, including logistics and transport.
FIDEUS	6th FP	Project whose aim is to develop innovative vehicle solutions for urban deliveries as well as a telematics platform to support the organisation of deliveries and transhipment points, and to provide a tool for local authorities to be able to monitor and manage delivery traffic more efficiently.
LEAN	4th FP – TRANSPORT	Development and demonstration of new concepts, telematics applications and administrational tools to distribute and collect goods in urban areas. The project included alternative transport mode recommendations to support significant modal shift to rail.
MOSCA	5th FP – IST	Key project objective is to provide a set of tools for improving efficiency of door-to-door transport of goods in urban areas by collaboratively providing demand and supply side information in one single environment/system.
PARCELCALL	5th FP – IST	Development and verification of an open architecture for intelligent tracking and tracing in transport and logistics.
THEMIS	5th FP -TREN	Examination of the potential for integrating traffic management and fleet management systems across all the modes, including an investigation of the implications for ITS architectures.

7. ITS SUPPORT

7.1 Architecture		
COMETA	4th FP - DG INFSO	Development of an electronic and telematics architecture for commercial vehicles.
COMPRIS	5th FP -TREN	Enhancement of the existing River Information Services (RIS) concepts to improve the safety of inland shipping traffic, environmental protection, the optimisation of inland port resources and the management of traffic flows.
FRAME-NET	5th FP - IST	Provision of information about the European ITS Architecture and promotion of opportunities for those involved in architecture-related activities in Europe to exchange their experience.
FRAME-S	5th FP -IST	'Accompanying measure' of FRAME-NET to update the European ITS Framework Architecture, creation and running of training seminars and workshops, technical assistance and development of navigation tools for the framework architecture.
GERDIEN	3rd FP – DRIVE 2	Architectural project to develop an open framework for traffic data collection and exchange the flexible integration of Advanced Transport Telematics (ATT) applications at all levels. It will cover a major part of the infrastructure required for dynamic traffic management in interurban parts of the Integrated Road Transport Environment (IRTE).
KAREN	4th FP – IST	Project to allow the complete process to be followed from the establishment of European requirements, through the production of a comprehensive European Transport Telematics Architectures Framework, to the build-up of a consensus and concluding with the endorsement of its results.
QUARTET	3rd FP – DRIVE 2	Development of a fully Integrated Road Transport Environment (IRTE) based on five integrated Advanced Transport Telematics (ATT) technologies – target IRTE architectures, Environment Control, Dual Mode Route Guidance, Public Transport Management and Information Systems, and Emergency Call.
THEMIS	5th FP -TREN	Thematic Network for Intermodal Freight Transport Information and Management Services. It is focused on the integration of Traffic Management systems with Intermodal Freight Information systems.

7.2 Radio Navigation		
ADASE 2	5th FP – IST	About 30 projects organised in the ADASE cluster, doing research for driver assistance systems and related activities to improve road transport safety, efficiency and comfort.
GALILEO	EC	European Satellite Navigation System. The system will provide quality services in all environments: cities, indoors and outdoors, land, sea, air, etc. in synergy with other service provisions, such as telecommunication, surveillance or observation.

7.3 Education and Training		
Asia-Pacific Academic Network	Japan	Network of Excellence in the Asia-Pacific region for ITS training.
ETNITE	6th FP Leonardo da Vinci	European Network on ITS Training and Education to support the ITS EduNet, which will offer validated training material in order to enhance the technical skills, knowledge and general awareness of ITS.
FRAME-S	5th FP - IST	'Accompanying measure' of FRAME-NET to update the European ITS Framework Architecture, creation and running of training seminars and workshops, technical assistance and development of navigation tools for the framework architecture.
HUMANIST	6th FP	Network of Excellence to create a virtual research centre concerning human-related design with IT technologies applied to road transport.
PCB	USA DoT	To provide information for anyone interested in improving their knowledge of ITS, at various levels of specialisation.
PORTAL	5th FP -TREN	Aimed to accelerate the take-up of EU research results on urban and regional transport, through the development of new teaching material, translated into 16 languages of the EU.

Appendix B

Acronyms and Abbreviations

AAM	Alliance of Automobile Manufacturers
ACC	Adaptive/Autonomous Cruise Control
ADAS	Advanced Driver Assistance Systems
AM/FM	Amplitude Modulated/Frequency Modulated
ANPR	Automatic Number Plate Reading
ATT	Advanced Transport Telematics
AVI	Automatic Vehicle Identification
AVL	Automatic Vehicle Location
AVM	Automatic Vehicle Management
B2B	Business to Business
B2C	Business to Customer
CCTV	Closed Circuit Television
CDGPS	Carrier-phase Differential Global Positioning System
CEN TC 278	Comité Européen de Normalisation – Technical Committee , 278 = Road Traffic and Transport Telematics
CEPS	Common Electronic Purse Specifications
CEUs	Continuing Education Units
CGI	Common Gateway Interface
CIS	Commonwealth of Independent States
CITE	Consortium for ITS Training and Education
CLI	Caller Line Identification
CNT	Conseil National des Transports
CO	Carbon Monoxide
CO_2	Carbon Dioxide
CRM	Customer Relationship Management
CS	Commercial Services
CVHS	Co-operative Vehicle Highway Systems
DAB	Digital Audio Broadcasting
DAS	Driver Assistance Systems
DATEX	Data Exchange
DGNSS	Differential Global Navigation Satellite System
dGPS	Differential Global Positioning System
DG TREN	Directorate-General Energy and Transport

DIS	Driver Information Systems
DOT	Department of Transportation
DRT	Demand Responsive Transport
DRTS	Demand Responsive Transport System
DSRC	Dedicated Short Range Communication
EC	European Commission
ECTRI	European Conference of Transport Research Institutes
EDI	Electronic Data Interchange
EGNOS	European Geostationary Navigation Overlay Service
EMS	Emergency Medical Services
EOTD	Enhanced Observed Time Difference
ERTMS	European Rail Traffic Management System
ESoP	European Statement of Principles
ETA	Expected time of arrival
ETCS	European Train Control System
ETSI	European Telecommunications Standards Institute
EU	European Union
FCD	Floating Car Data
FDE	Fault Detection and Exclusion
FM	Frequency Modulated
FP	Framework Programme
GDP	Gross Domestic Product
GIS	Geographic Information System
GNSS	Global Navigation Satellite System
GPRS	General Package Radio System
GPS	Global Positioning System
GSM	Global System for Mobile communication
HC	Hydrocarbon
HGV	Heavy Goods Vehicle
HMI	Human-Machine Interface
HTML	Hypertext Markup Language
ICT	Information and Communication Technology
ID	Identification
IFTS	Intelligent Freight Transport Systems
IGT	Intelligent Grouping Transportation
IMU	Inertial Measurement Unit
IPSEC	Internet Protocol Security
IRTE	Integrated Road Transport Environment
IST	Information Society Technologies
ITS	Intelligent Transport System

IVI	Intelligent Vehicle Initiative
IVIS	In-vehicle Information Systems
IVS	In-vehicle system
IVT	In-Vehicle Terminal
JAMA	Japanese Automobile Manufacturers Association
LDWA	Lane Departure Warning Assistance
LED	Light Emitting Diode
LIF	Location Inter-operability Forum
LORAN	Long-Range Navigation
MLP	Mobile Location Protocol
MJM	Multi-jurisdictional Mayday
MMS	Multimedia Messaging Service
MoU	Memorandum of Understanding
MSAS	Multi-functional Satellite Augmentation System
MSc	Master of Science
MSD	Minimum set of data
MTSAT	Multi-functional Transport Satellite
NELS	North West European LORAN-C System
NHTSA	National Highway Traffic Safety Administration
NoE	Network of Excellence
NPA	National Police Agency
NPS	New propulsion technologies and systems
OBU	Onboard Unit
O-D	Origin-Destination
OECD	Organisation for Economic Co-operation and Development
OS	Open Services
OTDOA	Observed Time Difference of Arrival
PC	Personal Computer
PCB	Professional Capacity Building
PCMCIA	Personal Computer Memory Card International Association
PDA	Personal Digital Assistant
PhD	Doctor of Philosophy
POI	Point of Interest
PPP	Public Private Partnership
PRN	Private Radio Network
PRS	Public Regulated Services
PSAP	Public Service Answering Point
PT	Public Transport
PTA	Personal Travel Assistant
PTO	Public Transport Operations

PTS	Passenger Transport Services / Public Transport Systems
PuSHMe	Puget Sound Help Me
R&D	Research and Development
RAIM	Receiver Autonomous Integrity Monitoring
RDS/TMC	Radio Data System/Traffic Message Channel
RFID	Radio Frequency Identification Devices
RTD	Research and Technical Development
RTK	Real Time Kinematics
RUC	Road User Charging
SA	Selective Availability
SAR	Search and Rescue
SBAS	Satellite Based Augmentation Services
SELS	Southern European LORAN-C System
SME	Small and Medium Enterprise
SMS	Short Message Service
SOAP	Simple Object Access Protocol
SOL	Safety of Life
SoP	Statement of Principles
TAP	Telematics Application Programme
TCP/IP	Transmission Control Protocol/Internet Protocol
TDC	Travel Dispatch Centre
TEC	Traffic Engineering and Control
TEMPO	Trans-European intelligent transport systeMs PrOjects
TERN	Trans European Road Network
TIC	Traffic Information Centre
TIC	Traffic Information Consortium
TPEG	Transport Protocol Experts Group
TQM	Total Quality Management
TTI	Traffic and Traveller Information
TUC	Traffic-responsive Urban Control
UMTS	Universal Mobile Telecommunications System
USDOT	US Department of Transportation
UTC	Urban Traffic Control
UTMS	Urban Traffic Management Systems
VASP	Value Added Service Provider
VII	Vehicle Infrastructure Integration
VMS	Variable Message Sign
VP	Value Pricing
VSCS	Vehicle Scheduling and Control System
WAAS	Wide Area Augmentation System

WAP	Wireless Application Protocol
WiFi	Wireless Fidelity
WIM	Weigh-in-motion
XFCD	Extended Floating Car Data
XML	Extensible Markup Language

Bibliography

The bibliography is divided into the respective chapters.

2. CONTEXT

Banister, D. (1999): *Planning more to travel less: Land use and Transport.* Town Planning Review 70 (3) pp. 313-338

DETR (2001): *Older People: Their Transport Needs and Requirements – Summary Report*, http://www.mobility-unit.detr.gov.uk/older/needs/04.htm#1

European Commission (2001): (1) *White paper: European transport policy for 2010: time to decide*. COM (2001) 370, www.europa.eu.int

European Commission (2001): (2) *The institutions and bodies of the European Union: who's who in the European Union? What difference will the Treaty of Nice make?* www.europa.eu.int

European Commission (2003): (1) *The regular report*, www.europa.eu.int

European Commission (2003): (2) *Transport infrastructure charging policy*, www.europa.eu.int

European Commission (2003): (3) *Europe at a crossroads: the need for sustainable transport*, European Commission, Directorate-General for Press and Communication.

European Commission (2003): (4) *How the European Union works: a citizen's guide to the EU institutions*, www.europa.eu/int/comm/publications

European Commission and Eurostat (2004): *European Union energy & transport in figures*, EC DG-TREN

European Commission Transport website. http://europa/pol/trans/index-en.htm

Fontaine, P. (2003): *Europe in 12 lessons*, Booklet for the European Commission, www.europa.eu.int

Glaister, S. and Graham, D. (2003): *Transport pricing and investment in England: technical report*, Report for the Independent Transport Commission, www.trg.soton.ac.uk/itc

Glaister, S. and Ochieng, W. (2003): *Maximum mobility: the smart answer*, Paper to the Royal Society of Arts, Manufactures and Commerce.

ILS NRW; Babtie spol.; Langzaam Verkeer and ETT (2005): *Towards passenger intermodality in the EU*. Report for the European Commission, DG Energy and Transport.

Lynham, B. (1997): *Traffic and Health*. European Federation for Transport and the Environment, 97/7.

McDonald, M. et al. (2000): *The EC Telematics Applications Programme Activities for ITS in Cities*. Transportation Research Group, University of Southampton.

OECD (2002): *Project on environmentally sustainable transport,* Working Group on National Environmental Policy, Working Group on Transport. RNV/EPOC/WPNEP/T(2001)6/FINAL

ROSETTA (2004): (1) *Passenger transport services*, D9: Work Areas Final Report for DG IST of the European Commission, www.trg.soton.ac.uk/rosetta

ROSETTA (2004): (2) *Emergency service systems*, D9: Work Areas Final Report for DG IST of the European Commission, www.trg.soton.ac.uk/rosetta

ROSETTA (2004): (3) *Human machine interaction*, D9: Work Areas Final Report for DG IST of the European Commission, www.trg.soton.ac.uk/rosetta

ROSETTA (2004): (4) *Safety – enforcement*, D9: Work Areas Final Report for DG IST of the European Commission, www.trg.soton.ac.uk/rosetta

ROSETTA (2004): (5) *Traffic control and Road user charging*, D9: Work Areas Final Report for DG IST of the European Commission, www.trg.soton.ac.uk/rosetta

ROSETTA (2004): (6) *Multiple monitoring of road traffic*, D9: Work Areas Final Report for DG IST of the European Commission, www.trg.soton.ac.uk/rosetta

ROSETTA (2004): (7) *Personal travel services*, D9: Work Areas Final Report for DG IST of the European Commission, www.trg.soton.ac.uk/rosetta

ROSETTA (2004): (8) *Vehicles and highway ITS infrastructure*, D9: Work Areas Final Report for DG IST of the European Commission, www.trg.soton.ac.uk/rosetta

ROSETTA (2004): (9) *Radio navigation*, D9: Work Areas Final Report for DG IST of the European Commission, www.trg.soton.ac.uk/rosetta

ROSETTA (2004): (10) *Freight transport services*, D9: Work Areas Final Report for DG IST of the European Commission, www.trg.soton.ac.uk/rosetta

ROSETTA (2004): (11) *Training and education for ITS*, D9: Work Areas Final Report for DG IST of the European Commission, www.trg.soton.ac.uk/rosetta

ROSETTA (2004): (12) *ITS architecture,* D9: Work Areas Final Report for DG IST of the European Commission, www.trg.soton.ac.uk/rosetta

Stevens, A. and Stevens, H. (2000): *Brussels bureaucrats?* Palgrave ISBN 0333604903

Stevens, H. (2003): *Transport policy in the European Union*, Palgrave ISBN 0333793560

Yrjo-Koskinen, E. (2002): *Challenge to EU transport policy*, Available at http://www.sll.fi/tiedotus/tiedotteet/pressrelease/eykEUpolicy

3. TRAVELLER SERVICES

CORDIS – Community Research & Development Information Service, Luxembourg. Office for Official Publications of the European Communities (2003): Information Society Technologies for Transport and Mobility, Luxembourg

ERTICO booklet (2002), The Vision.

European Commission (2001): (1) *White paper: European transport policy for 2010: time to decide*. COM (2001) 370, www.europa.eu.int

European Commission (2001): (2) *Development of a legal and business framework for participation of the private sector in deploying telematics-based traffic and travel information TTI services in Europe*, 2001/551/EC

European Commission (2002): *Traffic and Traveller Information Services in Europe: Country Digests*

Kyoto Protocol to the United Nations Framework Convention on Climate Change (1997)

Picco, A. (2002): *Public Transport and use of ITS*, Internal Report of 5th FP project VOYAGER, WG 5

Salon, D., Sperling, D., Shaheen, S. and Sturges, D. (1999): *New Mobility – Using Technology and Partnership to Create More Efficient, Equitable, and Environmentally Sound Transportation*, University of California-Davis (UCD-ITS-RR-99-1)

Verband Deutscher Verkehrsunternehmen (Ed.) (2001): *Telematics in Public Transport in Germany*, Düsseldorf, Germany

4. VEHICLES AND INFRASTRUCTURE

Alkim, T. (2003): *Field Operational Test with Lane Departure Warning Assistant systems - Behavioural effects*. 10th World Congress on Intelligent Transport Systems. 16-20 Nov. 2003, Madrid, Spain

Alkim, T., Bootsma, E., Berghout, G., Ostyn and Gendre, P. (2004): *Adase 2 Expert Workshop on effects of ADA systems on safety, throughput and comfort*, ADAS 2 Deliverable D8E, contract number IST-2000-28010, Jul. 2004

Alexander, L., Cheng, P. M., Donath, M., Gorjestani, A., Newstrom, B., Shankwitz, C. and Trach, W. Jr. (2005): *Bus Rapid Transit Technologies: Assisting Drivers Operating Buses On Road Shoulders, Volume 1*. University of Minnesota, Center for Transportation Studies, 511 Washington Avenue, SE, Minneapolis, MN, 55455-0375, USA. Research report: CTS 04-12, Report Number: Final Report

Arc Website: www.arcgroup.com/automotive. Last accessed 12 Jul. 2004

Augello, D.J. (1991): *Description of three PROMETHEUS demonstrators having potential safety effects*. The Thirteenth International Technical Conference on Experimental Safety Vehicles, 4-7 Nov., 1991, PARIS, pp. 209-212

Biding, T. (2004): *ISA-Deployment -Demonstration towards Implementation.* 11th World Congress: on Intelligent Transport Systems, Oct. 18-24, Nagoya, Japan

Bishop, R. (2001): *Co-operative Intelligent Vehicle-Highway Systems: Status of Activities Worldwide, Stakeholder Perspectives, and Major Trends.* Conference Title: 8th World Congress on Intelligent Transport Systems. 30 Sept.–4 Oct. 2001, Sydney, Australia

Bonnet, C. and Fritz, H. (2000): *Fuel Consumption Reduction Experienced by two PROMOTE-CHAUFFEUR Trucks in Electronic Towbar Operation.* Proceedings of The 7th World Congress On Intelligent Systems, Held Turin, Italy, 6-9 Nov. 2000

CGALIES (2002): *Final report on implementation issues related to access to location information by emergency services (e-112) in the European Union.* www.cgalies.telefiles.de

Crawford, D. (2003): *Transport's pilot. Surveyor.* 190(5734), 30-31

ERTICO: *Expected Benefits of ITS*, 4th World Congress of ITS, (1997), Berlin, Germany

ESafety website: http://europa.eu.int/information_society/programmes/esafety/index_en.htm. Last accessed 12 Jul. 2004

European Commission (2000): Recommendation of 21st December 1999. On safe and efficient in-vehicle information and communication systems: *A European statement of principles on human-machine interface.* Official Journal, 25 Jan. 2000. 2000/53/EC

European Commission (2002). Final report of the eSafety Working Group on Road Safety. www.europa.eu.int

Hanowski, R. J., Carroll, R. J., Dingus, T. A. and Neale, V. L. (2002): *The Application Of State-of-the-Art Instrumented Vehicles to Driving Research and Its IVI Technology Assessment.* Conference Title: 9th World Congress on Intelligent Transport Systems. 14-17 Oct. 2002, Chicago, Illinois

Hou To, Choudry, O. (2000): *Mayday Plus Operational Test, Evaluation report.* Minnesota Guidestar Minnesota DOT, Apr. 2000

Hu, Z., Uchimura, K. and Lamosa, F. (2004): *Towards a New Generation of Car Navigation System - Data Fusion Technology in Solving On-Board Camera Registration Problem.* 11th World Congress on Intelligent Transport Systems, Oct. 18-24, Nagoya, Japan

Kwon, E., Kim, S. and Betts, R (2003): *Route-Based Dynamic Preemption of Traffic Signals for Emergency Vehicle Operations.* National Research Council (U.S.). Transportation Research Board. Meeting (82nd : 2003 : Washington, D.C.)

MacNeille, P. and Miller, R. (2004): *Co-operative Adaptive Cruise Control for Improved Mobility and Safety.* Conference Title: At the Crossroads: Integrating Mobility Safety and Security. ITS America 2004, 14th Annual Meeting and Exposition. 26-28 Apr. 2004, San Antonio, Texas

Malone, K. M., Eijkelenberg, P. (2004): *A Quick Scan of Quantified Effects of Advanced Driver Support Systems.* TNO 04-7N-192-74070, 2004

Marsden, G., Mcdonald, M. and Brackstone, M. (2001): *Towards an understanding of adaptive cruise control.* Transportation Research, Part C, 9C(1), 33-51

Nico, K. and Klaus, C. J. D. (2004): *Fusion of Laserscanner and Video for Advanced Driver Assistance Systems.* 11th World Congress on Intelligent Transport Systems, Oct. 18-24, Nagoya, Japan

Paola, C., Silvia, Z., Luisa, A. and Javier, F. de V. B. (2003): *AIDER – Vehicle and infrastructure equipment for emergency call management efficiency improvement.* 10th World Congress on Intelligent Transport Systems, 16-20 Nov. 2003, Madrid, Spain

Parent, M. and Gallais, G. (2003): *Cybercars : Review of First Projects.* 9th International Conference on People Mover. Sept. 2003, Singapore

Parent, M. and Gallais, G, (2002): *Intelligent Transportation in Cities with CTS.* ITS World Congress. Oct. 2002, Chicago, Illinois

Siuru, B. (2004:) PHILEAS: *A New Idea in Public Transport. MASS TRANSIT.* 2004/03. 25(2), pp. 44-46

Sparmann, J. M. (1991): *Acceptance and Benefits of The Berlin Route Guidance and Information System (LISB).* Conference Title: Proceedings of the Conference, Strategic Highway Research Program and Traffic Safety on Two Continents. Location: Gothenburg, 18-20 Sep. 1999, Sweden

Stefan, M. (2003): *Technology options for Intelligent Speed Adaptation.* 10th World Congress on Intelligent Transport Systems. 16-20 Nov. 2003, Madrid, Spain

Takuya, N. (2003): *An abnormal vehicle behaviour detection system by image processing in urban roadway tunnels.* 10th World Congress on Intelligent Transport Systems. 16-20 Nov. 2003, Madrid, Spain

Underwood, G., Jiang, C. and Howarth, C. I. (1993): *Modelling of safety measure effects and risk compensation. Accident Analysis and Prevention*, 25, pp. 277-288

U.S. Department of Transportation (2005): *Vehicle Infrastructure Integration (VII)* at http://www.its.dot.gov/vii/

Vöge, T. and McDonald, M. (2004): *Analysis of Potentials and Limitations of Automated Urban Transport Systems.* 11th World Congress on Intelligent Transport Systems and Services, Oct. 2004, Nagoya, Japan. ISBN: 4-9902236-0-8

Yin, Y., Miller, M. A. and Shladover, S. E. (2004): *Assessment of the applicability of co-operative vehicle-highway automation systems to freight movement in Chicago*, Institute of Transportation Studies, University of California, 83rd TRB 2004

5. NETWORK MANAGEMENT

Arup, O. (2004): *Frameworks for Modelling the Variability of Highway Journey Times.* UK Department for Transport

Dingus, T. A., Hulse, M. C., Mollenhauer, M. A.; Fleischman, R. N., McGhehee, D. V. and Manakkal, N. (1997): *Effects of age, system experience and navigation technique on driving with an advanced traveller information system.* Human Factors vol. 39, pp. 177-199

CORDIS – Community Research & Development Information Service, Luxembourg. www.cordis.lu

Economist (2004): *The road tolls for thee*. The Economist Technology Quarterly. 12 Jun.

European Commission (1999): Final report on estimating transport costs. High level group on transport infrastructure charging. Available at www.europa.eu.int

European Commission (2001): *White paper: European transport policy for 2010: time to decide*. COM (2001) 370, www.europa.eu.int

European Commission (2003): (1) *Transport infrastructure charging policy*. Available at www.europa.eu.int

European Commission (2003): (2) Press release: charging for transport infrastructure use: the Commission proposes a revision of the Community framework. Available: www.europa.eu.int

European Commission (2003): (3) Proposal for a directive of the European Parliament and of the council: amending directive 1999/62/EC on the charging of heavy goods vehicles for the use of certain infrastructures. COM (2003) 448. Available at www.europa.eu.int

European Commission (2004): GALILEO: European satellite navigation system. Available at www.europa.eu.int

Glaister, S. and Graham, D. (2003): *Transport pricing and investment in England*: technical report. Report for the Independent Transport Commission available at www.trg.soton.ac.uk/itc

Gunn, H. (2001): *The value of small travel time savings: some new evidence*. Association of European Transport, European Transport Conference Seminar: Behavioural Modelling www.aetransport.co.uk

Hoose, N. (2003): *Personal Communication*

Independent Transport Commission (2003): *Road pricing: better for everyone*. www.trg.soton.ac.uk/trg

Jara-Diaz, S. (2001): *A close look at the value of leisure time travel savings*. Association of European Transport, European Transport Conference Seminar: Behavioural Modelling www.aetransport.co.uk

Keller, H. (2004): *Intelligent Transport Infrastructure for Trans European Networks ITS-TEN*. Position Paper of CORVETTE: Needs of ITS/ TEMPO for Sustainable TENs

Kulmala, R, Luoma, S., Stenberg U. (2003): VIKING Workshop – *Best practices on monitoring deployment*. 6-7 Nov. 2003, Stockholm, Sweden. Available at www.viking.tent-t.com

Lehman, B. J. and Crano, W. D. (2002): *The pervasive effects of vested interest on attitude-criterion consistency in Political judgement*. Journal of Experimental Social Psychology vol. 38, pp. 101-112

Linnett, S. (2004): *Beyond congestion charging*. Independent Transport Commission www.trg.soton/itc

PIARC (The World Road Association) (2004): The ITS handbook

Pister, K. (2004): *Smart Dust/Motes, Autonomous sensing and communication in a cubic millimetre*. Robotics and Intelligent Machines Laboratory, department of Electrical Engineering and Computer Sciences, University of California at Berkeley

Social Research Associates (2002): M60 JETTS: survey of local opinion. Unpublished report for the Government Office for the North West (some highlights of the overall study are at http://www.go-nw.gov.uk/transport/m60jJet1.pdf)

Ubbels, B. and Verhoef, E. (2003): *Behavioural responses to transport prices: a multidisciplinary perspective*. Paper to the 7th NECTAR conference, Sweden

Wardman, M., Whelan, G. A., Fowkes, A. S., Nellthorp, J., Bates, J. and Coombe, D. (2002): *UK value of time*. For the UK Department for Transport

6. GOODS TRANSPORT

Eurolog (Società per la Logistica Integrata e il Transporto Merci delle Ferrovie dello Stato)

European Commission (2001): *White paper: European transport policy for 2010: time to decide*. COM (2001) 370, www.europa.eu.int

Eurostat (European Statistical Office), website: http://epp.eurostat.ec.europa.eu

Léonardi, J., Baumgartner M. (2004): *CO_2 efficiency in road freight transportation: Status quo, measures and potential*. Transport Research Part D9, pp. 451-464

McKinnon, A. C. and Ge, Y. (2003): *Analysis of Transport Efficiency in the UK Food Supply Chain*. Logistics Research Centre, Heriot-Watt University, Edinburgh

OPTSE (2005): *Intermodal transport in Europe*. Bulletin of the Observatory on Transport Policies and Strategies in Europe, Conseil National des Transports, Double Issue Nos. 13-14, Apr. 2005

7. ITS SUPPORT

Arc Website: www.arcgroup.com/automotive. Last accessed 12 Jul. 2004

ESafety website: http://europa.eu.int/information_society/programmes/esafety/index_en.htm. Last accessed 12 Jul. 2004

European Commission (2000): (1) Recommendation of 21st December 1999. On safe and efficient in-vehicle information and communication systems: A European statement of principles on human-machine interface. Official Journal, 25 Jan. 2000. 2000/53/EC

European Commission (2000): (2) European ITS Framework Architecture – List of European ITS User Needs, D2.02 – Issue 1

European Commission (2004): Planning a Modern Transport System, A Guide to Intelligent Transport System Architecture: Why you need one and how to create it, Issue 2

Underwood, G., Jiang, C., and Howarth, C. I. (1993): *Modelling of safety measure effects and risk compensation.* Accident Analysis and Prevention, 25, pp. 277-288

U.S. Department of Transportation: National ITS Architecture. http://www.iteris.com/itsarch/